USING THE BUILDING REGULATIONS

Administrative Procedures

USING THE BUILDING REGULATIONS

Administrative Procedures

M J Billington

ELSEVIER
BUTTERWORTH
HEINEMANN

AMSTERDAM • BOSTON • HEIDELBERG • LONDON • OXFORD • NEW YORK • PARIS • SAN DIEGO •
SAN FRANCISCO • SINGAPORE • SYDNEY • TOKYO

Elsevier Butterworth Heinemann
Linacre House, Jordan Hill, Oxford OX2 8DP
30 Corporate Drive, Burlington, MA 01803

First published 2005
Copyright © 2005, Elsevier Ltd. All rights reserved

British Library Cataloguing in Publication Data
A catalogue record for this book is available from the British Library

Library of Congress Cataloguing in Publication Data
A catalogue record for this book is available from the Library of Congress

ISBN 0 7506 6257 3

For information on all Elsevier Butterworth-Heinemann
publications visit our website at www.books.elsevier.com

Typeset by Charon Tec Pvt. Ltd, Chennai, India
www.charontec.com
Printed and bound in United Kingdom

Working together to grow
libraries in developing countries

www.elsevier.com | www.bookaid.org | www.sabre.org

ELSEVIER   BOOK AID
International   Sabre Foundation

Contents

Preface

Unfortunately, it is a fact that the Building Regulations and Approved Documents get more complex with every update, often requiring the services of specialist professionals to make sense of the provisions (which even they find difficult to understand!). New areas of control are being introduced each year and the scope of the existing regulations is being extended with each revision.

The current Approved Documents only provide detailed guidance in the design of extremely simple and straightforward buildings using mainly traditional techniques. Larger and more complex buildings require access to other source documents and although these are referred to in the Approved Documents no details of their contents or advantages of use are given. Therefore, the Approved Documents are becoming less and less useful to anyone concerned with the design and construction of any but the most simple of buildings.

To date, currently published guidance works on the Building Regulations and Approved Documents have tended merely to restate the official guidance in a simpler fashion with additional illustrations. Some of them have restructured the guidance in an attempt to make it more user-friendly, but haven't really added value. Additionally, where the Approved Documents have made reference to alternative guidance sources the texts have tended to do the same – without attempting to state what advantages might be gained by using a different approach and without giving even the most basic details of what might be contained in the reference texts.

In early 2003, the author approached the publishers of this book, Elsevier, with an idea for a new kind of building control guidance publication. The aim would be to present a series of books, each one covering a separate Approved Document, which would provide users with much more detailed guidance than the Approved Documents currently provide. Furthermore, these would be written by experienced engineers, surveyors, architects and building control surveyors, etc., who had relevant specialist knowledge and hands-on experience.

This book is the first in the series and it was thought appropriate to start with the most neglected part – the building control system itself. Who (apart from staff working for one of the many building control bodies) has ever actually read the Building Regulations or the Building Act? Whilst many people know that they can use an Approved Inspector instead of the Local Authority for building control services how many people have an intimate knowledge of the Building (Approved Inspectors, etc.) Regulations? With insufficient knowledge of these legislative provisions it is possible that costly mistakes can be made which may result in a project or design being seriously compromised.

Part 1 of this book is introductory and gives a broad overview of the Building Regulations and the control system. It includes a brief historical account of the development of building control in England and Wales. This sets out the main determining factors which have led to the current system and is of more than purely academic interest. An understanding of these factors is vital if we are to understand the underlying reasons for many of the things we are asked to provide in our buildings by Government and building control bodies.

Part II covers the legal background and includes an in-depth analysis of the *Building Act 1984*, the *Building Regulations 2000* and the *Building (Approved Inspectors, etc.) Regulations 2000*. There are two Appendices (included in Part IV) which should be read in conjunction with this Part. These provide consolidated commentaries on the Regulations referred to above.

Part III deals with the administration of the Regulations. This includes the control systems that are available to an applicant when it is considered that Regulation compliance is needed; the enforcement actions that can be taken by building control bodies when it is alleged that work is not in compliance; the options open to an applicant to challenge the views of the building control body; and a discussion of the many ways in which an applicant may satisfy the technical requirements of the Regulations.

The aim of this book (and this series of books) is to provide a convenient, straightforward, comprehensive guide and reference to a complex and constantly changing subject. It must be stressed that the books in this series are a guide to the various Regulations and Approved and other documents, but are not a substitute for them. Furthermore, the guidance in the Approved and other source documents is not mandatory and differences of opinion can quite legitimately exist between controllers and developers or designers as to whether a particular detail in a building design does actually satisfy the mandatory functional requirements of the Building Regulations.

The intended readers are all those concerned with building work – architects and other designers, building control officers, approved inspectors, building surveyors, clerks of works, services engineers, contractors, and teachers in further and higher education, etc. – as well as their potential successors, the current generation of students.

The law in this book is stated on the basis of cases reported and other material available to us on 1 July 2005.

M.J. Billington

Acknowledgements

The author wishes to thank the Editorial and Production team at Elsevier for their help, patience and dedicated support in the publication of this book.

About this series of books

Whether we like it or not, the Building Regulations and their associated Government-approved guidance documents get more complex with every update, often requiring the services of specialist professionals (services engineers, fire engineers, etc.) to make sense of the provisions. New areas of control are being introduced each year and the scope of the existing regulations is being extended with each revision.

The technical guidance given in the current Approved Documents is only of use in the design of extremely simple and straightforward buildings using mainly traditional techniques. For larger and more complex buildings it is usually better (and more efficient in terms of building design) to use other sources of guidance (British and European Standards, Building Research Establishment Reports, etc.) and although a great many of these other source documents are referenced in the Approved Documents no details of their contents or advantages of use are given.

The current Approved Documents usually fail to provide sufficient guidance just when it is needed, that is when it is proposed to deviate from the simple solutions or attempt to design something slightly unusual, thus encouraging adherence to traditional and (perhaps) unimaginative designs and details, and discouraging innovation in the majority of building designs.

This series of books, by addressing different parts of the Building Regulations in separate volumes, will enable each Part to be explored in detail.

The information contained in the Approved Documents is expanded not only by describing the traditional approach but also by making extensive reference to other sources of guidance contained in them. These 'alternative approaches' (as they are called in the Approved Documents) are analysed and the most critical parts of them are presented in the text with indications of where they can be used to advantage (over the traditional approach).

As this is a new concept in building control publications our aim is to develop the series by including examples of radical design solutions that go beyond the Approved Document guidance but still comply with the Regulations. Such innovative buildings already exist, one example being the Queen's Building at De Montfort University in Leicester, which makes extensive use of passive stack ventilation instead of traditional opening windows or air conditioning.

About this book

This book presents a detailed analysis of the administrative and procedural regulations and the processes that can be used for building control. This is the most neglected part of the whole building control system and this is the first time that such an analysis has been presented in one publication.

It answers questions such as:

- Do the regulations apply to my building proposals?
- If the regulations apply, how can I ensure that what I am doing will comply?
- What alternative routes to compliance are available to me?
- How do I know which local authority to use?
- How do I go about finding an approved inspector?
- Would it be better for me to use a local authority or an approved inspector?
- Can a suitably qualified person certify my work?
- Do I need to pay fees for my building regulation approval?
- What can I do if the local authority refuses my application or says that my work does not comply?
- What can I do if the approved inspector says that my work does not comply?
- What can I do if I have carried out work without gaining building regulation approval?
- How do I prove that my work complies with the regulations?
- Where do I find guidance to help me comply with the regulations?
- What can the local authority do to make me correct my work if they claim that it does not comply and what can I do if I don't agree with them?

This book is aimed at designers, builders, students on construction and building surveying related courses, building control professionals and anyone else with an interest in the built environment. Its purpose is to keep them better informed and more able to deal with a complex and evolving area of law which directly affects everyone.

PART 1

Introduction

1

Series introduction

1.1 Introduction

Although we may not be aware of it, the influence of the Building Regulations is around us all the time. In our homes building regulations affect and control the:

- Size and method of construction of foundations, walls (both internal and external), floors, roofs and chimneys
- Size and position of stairs, room exits, corridors and external doors
- Number, position, size and form of construction of windows and external doors (including glazing)
- Methods for disposing of solid waste
- Design, construction and use of the services such as:
  - above and below ground foul drainage taking the waste from kitchen and bathroom appliances (including the design and siting of the appliances themselves)
  - rainwater disposal systems including gutters and downpipes from roofs, and drainage from paths and paving
  - electrical installations
  - heating and hot water installations using gas, oil or solid fuel
  - fire detection and alarm systems
  - mechanical ventilation systems.
- Design and construction of the paths outside the house that:
  - lead to the main entrance
  - are used to access the place where refuse is stored.

In a similar manner, they also affect the places where people go when away from their homes such as:

- factories, offices, warehouses, shops and multi-storey car parks
- schools, universities and colleges
- leisure, sport and recreation centres
- hospitals, clinics, doctors surgeries, health care centres and other health care premises

- hotels, motels, guest houses, boarding houses, hostels and halls of residence
- theatres, cinemas, concert halls and other entertainment buildings
- churches and other places of worship.

In fact, anything that can normally be considered to be a building will be affected by Building Regulations. But it is not just the design and construction of the building itself that is controlled.

The regulations also affect the site on which the building is placed in order to:

- lessen the effect of fire spread between neighbouring buildings
- permit access across the site for the fire brigade in the event of fire
- allow access for disabled people who may need to get from a parking place or site entrance to the building
- permit access for refuse collection.

1.2 What are the Building Regulations?

When asked this question most people (assuming that they have even heard of the regulations) will usually bring to mind a series of A4 documents with green and white covers, and the words 'Approved Document' on the front! Of course, these documents are not the Building Regulations, but have come to be regarded as such by most builders, designers and their clients, and it is this misconception that has led to a great deal of confusion regarding the true nature of the building control system and the regulations. When applied to England and Wales, the Building Regulations consist of a set of rules that can only be made by Parliament for a number of specific purposes. The purposes include:

- Ensuring the health, safety, welfare and convenience of persons in or about buildings and of others who may be affected by buildings or matters connected with buildings.
- Furthering the conservation of fuel and power.
- Preventing waste, undue consumption, misuse or contamination of water.

The regulations may be made 'with respect to the design and construction of buildings and the provision of services, fittings and equipment in or in connection with buildings'.

Originally (in Victorian times), the regulations (or byelaws as they were known then) were concerned only with public health and safety, but in the late twentieth century additional reasons for making building regulations were added so that it would now seem possible to include almost anything under the banner of 'welfare and convenience'. The Regulations are of two types:

(a) Those that deal with issues of procedure or administration such as:
- the types of work to which the regulations apply
- the method of making an application to ensure compliance and the information that must be supplied to the controlling authority

- the frequencies and stages at which the control authorities must be informed of the work
- details of the testing and sampling that may be carried out by the controlling authorities to confirm compliance
- what sorts of work might be exempted from regulation control
- what can be done in the event of the work not complying with the regulations.

(b) Those that describe the 'standards' which must be met by the building (called 'substantive' requirements) such as:
  - the ability of the building to:
    - retain its structural integrity
    - resist the effects of fire and allow people to escape if a fire should occur
    - resist dampness and the effects of condensation
    - resist the passage of sound
    - minimise the production of carbon dioxide by being energy efficient
    - be safe to use, especially where hazards of design or construction might exist, such as on stairways and landings or in the use of glass in windows, doors or as guarding
    - maintain a healthy internal environment by means of adequate ventilation.
  - the safe installation and use of the building's services including:
    - electric power and lighting
    - boilers, open fires, chimneys, hearths and flues
    - unvented heating and hot water systems
    - sanitary installations and above and below ground drainage
    - foul and waste disposal systems
    - mechanical ventilation and air conditioning systems
    - lifts and conveyors.

As the regulations are phrased in functional terms (i.e. they state what must be achieved without saying how this must be done) they contain no practical guidance regarding methods of compliance. The intention of this approach is that it gives designers and builders flexibility in the way they comply and it does not prevent the development and the use of innovative solutions, and new materials and methods of construction. Of course, much building work is done in traditional materials using standard solutions developed over many years and based on sound building practice. To assist designers and contractors in these accepted methods the Government has provided non-mandatory guidance principally in the form of 'Approved Documents', there being an Approved Document that deals with each substantive provision of the Building Regulations. This does not prevent the use of other 'official' documents, such as Harmonised Standards (British or European), and the adoption of other methods of demonstrating compliance such as past experience of successful use, test evidence, calculations, compliance with European Technical Approvals, the use of CE-marked materials, etc.

1.3 How are the Regulations administered?

For most types of building work (new build, extensions, alterations and some use changes) builders and developers are required by law to ensure that they comply with the Regulations. At present this must be demonstrated by means of an independent check that compliance has been sought and achieved.

For this purpose, building control is provided by two competing bodies: Local Authorities and Approved Inspectors.

Both building control bodies will charge for their services. They may offer advice before work is started, and both will check plans of the proposed work and carry out site inspections during the construction process to ensure compliance with the statutory requirements of the Building Regulations.

1.3.1 Local Authority building control

Each Local Authority in England and Wales (Unitary, District and London Boroughs in England and County, and County Borough Councils in Wales) has a building control section. The Local Authority has a general duty to see that building work complies with the Building Regulations unless it is formally under the control of an Approved Inspector.

Individual local authorities co-ordinate their services regionally and nationally (and provide a range of national approval schemes) via LABC Services. Full details of each local authority (contact details, geographical area covered, etc.) can be found at www.labc-services.co.uk.

1.3.2 Approved Inspectors

Approved Inspectors are companies or individuals authorised under sections 47 to 58 of the Building Act 1984 to carry out building control work in England and Wales.

The Construction Industry Council (CIC) is responsible for deciding all applications for Approved Inspector status. A list of Approved Inspectors can be viewed at the Association of Consultant Approved Inspectors (ACAI) web site at www.acai.org.uk. Full details of the administrative provisions for both Local Authorities and Approved Inspectors may be found in Chapter 5.

1.4 Why are the Building Regulations needed?

1.4.1 Control of public health and safety

The current system of building control by means of Government regulation has its roots in the mid-Victorian era. It was originally set up to counteract the truly horrific living and working conditions of the poor working classes who had flocked to the new industrial towns in the forlorn hope of making a better living. Chapter 2 describes the factors which caused this exodus from the countryside and the conditions experienced by the incomers; factors which led to overcrowding, desperate insanitary living conditions, and the rapid outbreak and spread of disease and infection. There is no doubt that a punitive system of control was needed at that time for the control of new housing, and the enforcement powers given to local authorities (coupled with legislation that dealt with existing sub-standard housing) enabled the worst conditions to be eradicated and the spread of disease to be substantially halted.

The Victorian system of control based purely on issues of public health and safety enforced by local authorities continued to be effective for the next 100 years, the only major change being the conversion of the system from local byelaws to national regulations in 1966.

1.4.2 Welfare and convenience and other controls

The first hint of an extension of the system from one based solely on public health and safety came with the passing of the Health and Safety at Work Act in 1974 (the 1974 Act). Part III of the 1974 Act was devoted entirely to changes in the building control system and regulations, and it increased the range of powers given to the Secretary of State. Section 61 of the 1974 Act enabled him to make regulations for the purposes of securing the welfare and convenience (in addition to health and safety) of persons in or about buildings. Regulations could also be made now for furthering the conservation of fuel and power, and for preventing the waste, undue consumption, misuse or contamination of water. The 1974 Act was later repealed and its main parts were subsumed into the Building Act 1984 (the 1984 Act, see Chapter 3).

1.4.3 The new system and the extension of control

Initially, the new powers remained largely unused and it was not until the coming into operation of the completely revamped building control system brought about by the 1984 Act and the Building Regulations 1985 that the old health and safety based approach began to change. The 1984 Act also permitted the building control system to be administered by private individuals and corporate (i.e. non-local authority) bodies called Approved Inspectors in competition with local authorities, although enforcement powers remained with local authorities. The new powers have resulted in the following major extensions of control to:

- heating, hot and cold water, mechanical ventilation and air conditioning systems
- airtightness of buildings
- prevention of leakage of oil storage systems

- protection of liquid petroleum gas (LPG) storage systems
- drainage of paths and paving
- access and facilities for disabled people in buildings
- provision of information on the operation and maintenance of services controlled under the regulations
- measures to alleviate the effects of flooding in buildings
- measures to reduce the transmission of sound within dwellings and between rooms used for residential purposes in buildings other than dwellings.

Furthermore, a recent consultation in 2004 put forward proposals intended to facilitate the distribution of electronic communication services (Broadband) around buildings in a proposed Part Q, presumably under the banner of convenience.

As the scope of control has increased, the Government has attempted to simplify the bureaucratic processes that this increase would undoubtedly lead to by allowing work of a minor nature and/or service installations to be certified as complying by a suitably qualified person (e.g. one who belongs to a particular trade body, professional institution or other approved body).

1.4.4 The future of building control in England and Wales

This section derives its title from a Government White Paper (Cmnd 8179) published in February 1981. In paragraph 2 of this document the Secretary of State set out the criteria which any new building control arrangements would be required to satisfy. These were:

- maximum self-regulation
- minimum Government interference
- total self-financing
- simplicity in operation.

One out of four (total self-financing) may not seem to be a particularly good result and it has often been the case that the average local authority building control officer has been inadequately prepared through inappropriate education and training to take on the task of assessing compliance with many of the regulation changes listed above. It has been claimed that this problem has been solved by the introduction of Approved Inspectors onto the building control scene. Since these are staffed almost entirely by ex-local authority building control officers, it would seem that the net result of the partial privatisation of building control has only been to redistribute a finite number of similar people without any improvement in education or training, although the adoption of a more commercial attitude by Approved Inspectors may be a good or bad thing depending on your point of view.

It seems almost inevitable (without a change of Government or in Government thinking) that the areas of control will increase and that more 'suitably qualified people' will be entitled to certify work as complying with the regulations. It is also likely that local authorities will remain as the final arbiter in matters of enforcement although it likely that their direct involvement in day-to-day building control matters will diminish, to be taken over by the private sector. Indeed, most building control work on new housing is already dealt with by the private sector.

Although the broad subject area covered by the Building Regulations is roughly the same across the European Union (EU) (and in former British colonies such as Canada, Australia and New Zealand) the main difference between the system in England and Wales, and that in other countries, lies in the administrative processes designed to ensure compliance. Our mix of control mechanisms encompassing both public (Local Authority) and private (Approved Inspector) building control bodies offers not only choice but also potential conflict. The system is further complicated by the existence of certain 'self-certification' schemes for the installation of, for example, replacement windows and doors or combustion appliances, and some work, which has to comply with the regulations, but is 'non-notifiable' if carried out by a suitably qualified person.

In fact, we are the only country in the EU with such a 'mixed economy'. Most countries (Scotland and Northern Ireland, Denmark, The Netherlands, the Irish Republic, etc.) use a system run exclusively by the local authority. In Sweden the building control system was privatised in 1995 so that the work of plan checking and site inspections is carried out by a suitably qualified 'quality control supervisor' employed by the building owner; although the local authority still has to be satisfied that the work is being properly supervised and may carry out spot checks and inspections to confirm this.

Some years ago the UK Government consulted on proposals to extend control of work governed by the Building Regulations to a range of bodies (some of which could be engaged in design and construction) provided that they were suitably qualified and insured. For example, this would mean, that a firm of architects would be able to take complete control of their own building control processes for work that they had designed without using a local authority or Approved Inspector. Such a market-led system would seem to be in accordance with the aims listed at the beginning of this section and, provided that the necessary safeguards could be put in place to prevent corruption and build public confidence, it would seem to be a sensible way forward. The consultation exercise did not, however, result in any companies being approved to control their own work, although the current system of certification of compliance by suitably qualified persons did come out of the exercise. Whether this was caused by political interference, objections from the building control establishment or lack of confidence by companies who still wanted the comfort of a third party to do their regulation checking for them, is not known.

Brief history of building control

2.1 Introduction

This chapter sets building control in England and Wales into its historical context. Although forms of building control have existed since the twelfth century, the first comprehensive Building Act was not placed on the statute book until after the great fire of London, which occurred in 1666. A number of towns in England also passed building control legislation in the early Victorian era (e.g. the Liverpool Building Act which came into force in about 1842), but it was not until 1875 that general powers to control construction by means of byelaws (i.e. local laws) was extended to the most parts of England and Wales. Byelaws (and later, the regulations) have been passed for a number of reasons. Initially, this was for the purposes of protecting the health and safety of the public after it was recognised that poor construction (e.g. the production of buildings that were structurally unsound, damp, unventilated, poorly lit, and lacking adequate sanitary facilities, clean water and refuse disposal) led to the spread of disease and caused general ill health. It was also recognised that places of work could be just as injurious to health and safety as poor living conditions, notably where the nature of the process being carried out could lead to an increased risk of fire (e.g. the cotton industry). Regulations have sometimes been introduced in response to disasters or the catastrophic failure of buildings (e.g. the collapse of Ronan Point in 1968). More recently, factors other than health and safety have been seen to be equally important, and regulation through the building control system is now seen as the natural channel for bringing in such improvements. Examples include the introduction of 'welfare and convenience' as reasons for making regulations (e.g. access and facilities for disabled people) and conservation of fuel and power (to prevent wastage of energy and ultimately, to have an effect on global warming). To some people it may seem that the Government is stretching the limits of what should be covered by the building control system. The obvious example is the proposed introduction of a new Part Q to facilitate the distribution of electronic communication services (Broadband) around buildings. Presumably this is being done under the banner of convenience.

2.2 The industrial revolution and the rapid development of towns

2.2.1 Birth of the industrial revolution

In 1709 Abraham Derby, a brass founder, discovered how to smelt iron using coke instead of the usual method using charcoal. The importance of this discovery cannot be underestimated since coke (which is made from coal) was readily available in large quantities, whereas charcoal did not lend itself to large-scale production and could only be produced where large quantities of wood was available. It took some time for the new technology to become widely known but it led ultimately to the rapid expansion of the iron industry in the 1750s. For large-scale production it was important at that time that the raw materials be found close together and there be good means of transportation. These conditions were found in the Ironbridge Gorge in Shropshire where it was possible to mine iron ore, coal, limestone, sand and clay, and where transport was readily available using the River Severn. This happy combination of factors led to the birth of the industrial revolution and Ironbridge can rightly be called its birthplace. Initially, blast furnaces were built on the sides of the Gorge adjacent to the river, where waterpower was used to operate the bellows which provided the draft for the furnaces, but it was only when James Watt's new steam engine was used to provide the draft in blast furnaces that the real change occurred. Blast furnaces and foundries moved from wooded regions to coal-producing areas and unlimited production became possible. From a production level of 17,000 tons in 1740, this had jumped to 650,000 tons by 1830.

2.2.2 Enclosure and traditional methods of production

Also at this time, technological and sociological changes were affecting the nature of the agricultural community. At the beginning of the eighteenth century the majority of the population was employed in agriculture or cottage industry, and a great deal of otherwise fertile land was deemed to be common land whereby agricultural workers were permitted certain commoners rights such as the grazing of animals or the collecting of fire wood. Reform of this system was driven by the passing of Enclosure Acts. A number of local landowners whose land bordered the common would join together (hopefully with the support of their local Member of Parliament, who in many cases was one of them) to propose a Parliamentary Bill with the objective of enclosing the common. Officially, the rights of the commoners had to be taken into account but this involved establishing that they had those rights in the first place. In a great many cases enclosure led to the ejection of commoners from their cottages and land, and the extinguishment of their rights leading to much hardship and even starvation.

Before the invention of the machines which allowed large-scale production methods, most traditional forms of manufacturing were based in the worker's homes. This is particularly true in the case of the production of cloth. The old family method of production in the worker's home where the thread was spun, woven and dyed, and from where the finished product was sold was structurally incapable of expansion. In order to keep up with the competition by increasing production and reducing costs, traders began to use groups of specialist workers. Further competition acted as a powerful spur to invention. The old and inefficient hand-loom was superseded by John Kay's invention of the 'flying shuttle' in 1773, which increased the speed of the weaver's operation. Hargreaves' 'spinning jenny', which was introduced in 1764, enabled a single worker to manage several threads and in 1771 R. Arkwright invented the 'water frame', the spinning machine worked by water which was replaced 8 years later by Crompton's 'mule', thus combining the benefit of all the earlier machines.

Furthermore, the invention of the steam engine in the late eighteenth century referred to above, revolutionised the whole pattern of the industry. In the new textile industries workshops moved to be near coalmines and to places where the damp climate suited the thread. For example, Lancashire became the main supplier of cotton goods to the world for a period 150 years, since the damp climate aided the production of the thread and the port of Liverpool together with the Mersey River provided a convenient means of transport, import and export. Lancashire imported 25,000 tons of raw cotton in 1800 and by 1861 this had risen to 300,000 tons.

2.2.3 Population growth and movement

During this time the growth in population was considerable. In 1801, the date of the first census, the population of England and Wales stood at 9 million, but in just 30 years this figure reached 14 million. During the same period the population of London, which was 900,000 in 1801, almost doubled. As the numbers grew, the economic changes that were taking place altered the distribution of the population completely. The flight from the countryside caused by the Enclosure Acts meant that the only alternative to agriculture was industrial work and this began to be concentrated increasingly on the centres of production. This resulted in the depletion of whole agricultural districts from the south of the country with families moving to the cramped districts near the factories and coalfields of the Black Country in Staffordshire, South Wales, South Yorkshire, Lancashire and Scotland. For example, a half-timbered village of 12,000 people in the middle of the eighteenth century, called Manchester, had jumped in population to 400,000 by 1850 and the populations of Glasgow and Leeds increased tenfold within 100 years.

2.2.4 Transportation and communications

With the rapid increase in production, means had to be found for transporting both the raw materials and the finished goods. Initially this resulted in an improvement of the road system so that after the middle years of the eighteenth century new turnpikes had replaced the rough parish roads. For example, it now only took 4½ days to get from Manchester to London using the 'flying coach' service! The road improvements continued with Metcalfe and Telford building new roads and bridges, and the invention of a new process of road making in 1810 by John Macadam. Additionally, a commercial experiment by the Duke of Bridgewater near Manchester in 1761 led to a boom in canals. Sea transport followed and communications further improved after Stephenson's locomotive had completed its first run between Stockton and Darlington in 1825, although it took the adoption of the standard gauge for the real impact of railways to be felt.

With the trade routes converging on the main towns these became the financial and administrative centres for the new methods of production which in turn revolutionised the use of land and changed the appearance of the countryside.

2.2.5 Political unrest and sub-standard housing

For most of the last 20 years of the eighteenth century and first 15 years of the nineteenth century England was at war with France, and intermittently with other countries including America, The Netherlands, Spain and Denmark. The Napoleonic Wars ended with the Battle of Waterloo in 1815 and the cessation of hostilities by land and sea resulted in large numbers of soldiers and seamen returning to their homes with little in the way of skills or a means to earn a living. Additionally, progressive inflation had arrived and between 1790 and 1813 the cost of living doubled. Many returning home hoped to take up former agricultural employment but found that the jobs no longer existed. To many the causes of their unemployment were the machines which could do the work more efficiently and with less use of labour than had previously been the case. The Luddite riots against the use of machines had reached their peak by 1820 and the disastrous protectionist Corn Laws had pushed up the price of bread. The Government's response was that deficiencies in wages should be made up by local rates (the Speenhamland system) but this had the opposite effect of sending wages down and it is not altogether surprising that the period reached the point of dreadful frustration and mutual suspicion culminating in Peterloo Massacre of 1819.

With virtually no controls over construction, the dwellings that were erected to house the industrial workers were small and of poor quality; and in many new centres of production, there were no representative local authorities to provide those services which are taken for granted today. Where the newer centres of production were distant from existing towns, mill and colliery owners built their own workers' housing for as little outlay as possible on cramped sites, since much of the best available land for building was required for the works. This led to the development of 'back-to-back' terraces with each house having only one external wall, resulting in poor natural lighting and ventilation. Since the right to occupy a house was linked to the need to work for a particular master, the rents of such 'tied' houses were often linked to the wages paid and were accordingly low. This meant that the capital cost of each house had to be kept to a minimum, which resulted in poor and cheap construction.

Inevitably, the increased pressure for accommodation caused by the influx of labour to the new industries meant that investors in housing development could make fantastic rates of return, since it was possible to extract a high proportion of a worker's wage by way of rent for even sub-standard accommodation.

When new, the houses were probably better than the country dwellings that the newcomers were used to, because they were constructed of slate and brick rather than wood and thatch. Additionally, they were used as dwellings only, instead of being littered with the machines of trade; however, there is no doubting their rudimentary nature. New problems were presented by the high density of population which had not existed in the rural communities. These included lack of proper sanitation and refuse collection, pollution from smoke and waste, lack of ventilation and natural lighting, and general congestion.

2.3 The period of reform

2.3.1 The Reform Act 1832

The Reform Act of 1832 spread political power over a much wider area by giving parliamentary representation to many areas inadequately represented before and led to a great deal of subsequent legislation. The Earl of Shaftsbury was the driving force behind new factory legislation which by stages was to abolish the worst aspects of working conditions (including the employment of children in the mines and as chimney sweeps) and in 1834 he was asked to look at reform of the poor law. One of the Commissioners under the new poor law legislation was Edwin Chadwick, the Former Assistant and friend of Jeremy Bentham, the reformer. In 1839 the first sanitary commission was appointed at Chadwick's instigation after a request for his help by the authorities in Whitechapel, London who were trying to deal with a local epidemic disease. The report of the Commissioners attracted wide attention, and it became a textbook of sanitation throughout the country. Chadwick was deeply concerned at the number of people admitted to the workhouses and realised that if the health of the working population could be improved then there would be a drop in the numbers of people on poor law relief. As a consequence, he embarked on a nation-wide investigation of public health resulting in the historic *Report on the Sanitary Condition of the Labouring Population of*

Great Britain which he published privately and at his own expense in 1842. The report included figures to show that in 1839 for every person who died of old age or violence, eight died of specific diseases. The Royal Commission on the Health of Towns was appointed in 1843 as a direct result of Chadwick's report. The Commission made a thorough investigation of the sanitary arrangements of 50 English towns. The recommendations of the Health of Towns Commission led to the passing of the Public Health Act 1848.

2.3.2 Epidemics and legislation: The Public Health Act 1848

In the 1830s and the 1840s there were three massive waves of contagious diseases. The first, from 1831 to 1833, included two influenza epidemics and the initial appearance of cholera. The second, from 1836 to 1842, encompassed major epidemics of influenza, typhus, typhoid and cholera. The third cholera outbreak from 1847 to 1848 coupled with the earlier epidemics probably forced the Government to take action.

The Public Health Act of 1848 was the first piece of legislation that attempted to deal with issues of public health. Initially, it established the General Board of Health for a period of 5 years with three commissioners, which was later increased to four. In 1854 the Board was reconstituted with a President responsible to Parliament. Localities could petition the Board for application of the Act in their areas (i.e. to form their own local boards of health). Where the average death rate exceeded 23 per 1000 the Board could create local boards on its own initiative. When a local board was created an inspection of the sanitary state of the area was made although no central inspection was required for authorities that had Boards of Health outside the legislation. Corporate boroughs were to assume responsibility for drainage, water supplies, removal of 'nuisances', paving, etc. Finance was to be raised from the rates to pay for improvements.

The Act was permissive rather than compulsory in towns other than Municipal Corporations; therefore, it suffered from the following weaknesses:

- Responsibility was optional and the General Board met resistance to its orders
- London, Scotland and Ireland were excluded from the Act
- No measures were taken to ensure professionalism
- The General Board had few powers once a local board was set up
- The General Board had no money
- The problem of public health was not made a ministerial responsibility.

The General Board of Health was abolished by the Public Health Act 1858, and its functions and staff were transferred in part to the Local Government Act Office, a sub-department of the Home Office, and in part to the Medical Department of the Privy Council. The Local Government Act Office dealt with correspondence and reports directed to the Home Secretary or its own secretary by local boards of health. Its three engineering inspectors also undertook investigations resulting from the application of the Sanitary Acts of 1866–1870, which obliged local authorities to inspect their districts and suppress nuisances.

By 1870 there were over 700 authorities working under the public health and local government legislation. There were also poor law and registration authorities operating within their boundaries, but uncoordinated. In small towns and rural areas parish vestries, boards of guardians, highways boards and other bodies all had a hand in public health matters. The report of Royal Sanitary Commission, set up in 1869 to investigate this complicated situation, resulted in further legislation: the Local Government Board Act 1871, which transferred to the Local Government Board the sanitary and public health functions of the Privy Council Medical Department; and two Public Health Acts 1872 and 1875, which established rural and urban sanitary authorities and made compulsory the appointment of medical officers of health to advise them.

2.3.3 The Public Health Act 1875

The Public Health Act of 1875 set up a completely new basis for the layout of houses and streets in England and Wales and by the time that the Public Health Act of 1891 had been passed a vast pattern of health legislation was in existence which set the pattern for future house building. Sanitary rules for new houses were stringent and exact, and the appointment of medical officers of health had been made in all districts to ensure that they were carried out. The Public Health legislation and the Artisans Dwellings Act of 1875, which permitted local authorities to purchase slum dwellings and replace them with modern, healthy housing, enabled many improvements to be made. One of the most significant results of the passing of the Public Health Act 1875 was that general powers to control construction by means of byelaws came into effect. Local authorities were empowered to make byelaws aimed at securing the interests of health by ensuring that:

- there was an adequate water supply, drainage and sewage disposal
- sufficient air space was provided about buildings and ventilation was provided to those buildings
- nuisances were to be removed
- offensive trades were regulated
- contaminated food was found, confiscated and destroyed
- cases of infectious diseases were reported to the local medical officer of health who then had to take appropriate action.

Further regulations dealt with matters concerning markets, streetlighting, slaughterhouses and burials.

The power to make byelaws was extended by an amending Act in 1890 to include the provision for the flushing of water closets, the height of rooms intended for human habitation, the paving of yards and open spaces around dwellings, and the provision of backyards intended to facilitate the removal of refuse from a dwelling.

Model byelaws were issued by the then Ministry of Health, and byelaws made by councils were normally based on this model. They were not operative until confirmed by the Ministry of Health. The model byelaws were offered as a guide only, and they were considerably varied in many towns, leading to a lack of uniformity.

Later legislation in 1888 set up county councils and enabled them to appoint medical officers of health and in 1894 established urban and rural district councils as health administrations.

Thus the basis was established for a local authority system of building control based on local byelaws, a system that remained largely unchanged in principle until the reforms of the 1960s.

2.3.4 The Public Health Act 1936 and the Model Byelaws

The Public Health Act 1936 repealed the greater part of earlier legislation, and provided for byelaws with respect to buildings and sanitation. But local authorities were not obliged to make byelaws, since the 1936 Act required that: 'every local authority may, and if required by the Minister, shall make byelaws for regulating all or any of the following matters':

- the construction of buildings and the materials to be used
- space about buildings, natural lighting and ventilation of buildings, and the dimensions of rooms intended for human habitation
- the height of buildings, the height of chimneys above roofs
- sanitary conveniences, drainage, and the conveyance of water from roofs and yards
- cesspools and other means for the reception or disposal of foul matter
- ashpits
- wells, tanks and cisterns for the supply of water for human consumption
- stoves and other fittings insofar as required for the purposes of health and the prevention of fire
- private sewers; junctions between drains and sewers, and between sewers.

It should be noted that the 1936 Act did not specifically cover sound insulation, thermal insulation, safety in the use of stairs and ramps, disability issues or means of escape in case of fire. It was concerned primarily with health and safety (welfare and convenience did not appear until the Health and Safety at Work etc Act 1974).

Byelaws made under the 1936 Act could also make provisions requiring:

- the deposit of plans and the giving of notices
- the inspection of work
- testing of drains and sewers, and the taking of samples of materials.

The byelaws also related to structural alterations or extensions of buildings and instances where any material change of use had taken place.

As explained above, although guidance from central government was given, local authorities still held direct responsibility for issuing their own byelaws. This led to significant differences between local authority areas, and made it necessary for designers to obtain and be familiar with each local set of byelaws. The situation was improved with the issue by the Government of the 1952 Model Byelaws, series IV (amended 1953). The new Model Byelaws were universally adopted throughout England and Wales (with the exception of London); however, local authorities were still able to make their own byelaws and these were usually appended at the end of the Model Byelaws. Additionally, a different technique of control was introduced whereby standards of performance were stated, and in many cases these formed the mandatory part of the byelaws. For example, descriptions of actual structural minima, which previously had been mandatory (e.g. minimum wall thicknesses, etc.), were now contained in the so-called 'deemed to satisfy provisions'. This enabled designers to adopt newer methods and materials, provided that their performance could be established. This system enabled increasing reference to be made to advisory publications such as British Standard Specifications and Codes of Practice. The Model Byelaws also contained major new controls introducing, for the first time, the concept of structural fire-resistance periods in relation to building type and size.

2.4 The Modern Era

2.4.1 Résumé

The consolidation of the control system set up under the *Public Health Act 1875* empowered local authorities to make and enforce building byelaws for their districts. National model byelaws were published, their adoption being voluntary, and most local authorities embraced the model byelaws whilst adding their own local variations. Since the byelaws were administered by more than 1400 local authorities, considerable variations existed over relatively small regional areas.

Adjustments to the system brought about by the *Public Health Act 1936* did little to promote uniformity of control. Eventually, this was dealt with by the *Public Health Act 1961*, which contained the powers to make National Building Regulations for England and Wales (although Scotland and Northern Ireland were not included in the reforms and Inner London was not included until 1987).

2.4.2 The Public Health Act 1961 and the first National Building Regulations

By the late 1950s it was becoming apparent that the existing system of local byelaws was leading to lack of uniformity and was incapable of responding to changes in building techniques and materials, and advances in knowledge. In an effort to remedy the situation, powers were taken in the Public Health Act 1961 to make National Building Regulations.

This effectively removed powers which the local authorities had possessed since the 1870s, to make building byelaws. The new procedure was for the Minister to make building regulations by way of the statutory instrument, having a universal application throughout England and Wales (excluding London). Initially, the regulations were to be made broadly to cover the list of objectives set out in the 1936 Act and local authorities were to remain responsible for enforcement of the new building regulations.

The 1961 Act introduced a new provision which enabled a local authority to dispense with or relax any requirement of the Building Regulations, where the authority (or the Minister) considered that the operation of the regulation would be unreasonable. When a local authority refused an application to relax or dispense with a requirement the applicant could, within a month, appeal to the Minister.

With advice from the Building Regulations Advisory Committee (which was set up under the 1961 Act) the first National Regulations for England and Wales (the *Building Regulations 1965*) were laid before Parliament on 22 July 1965 and came into force on 1 February 1966.

By today's standards the *Building Regulations 1965* were relatively simple to understand. They were also rather restrictive and relied heavily on deemed-to-satisfy provisions, so much so that these came to be regarded almost as mandatory technical rules. The 1965 Regulations contained a number of provisions which are now regarded as anachronistic. For example, there were controls over the height of rooms in dwellings, requirements for zones of open space to be provided outside the windows of habitable rooms, provision for the ventilation of larders and provisions for the design and construction of ashpits, wells, tanks, cisterns and earth-closets. Interestingly, comprehensive regulations covering conservation of fuel and power were not yet being considered and although there were regulations dealing with thermal insulation these applied only to dwellings. As an example, the standard for insulating the roof of a dwelling could be met using a 25-millimetre glass or rock fibre quilt! There were no provisions in the 1965 Regulations for means of escape in case of fire or for access and facilities for disabled people. Additionally, the 1965 Regulations were in Imperial units.

The Regulations passed through numerous amendments and consolidations (including metrication in the 1972 consolidation) and eventually became the *Building Regulations 1976 (SI 1976/1676)* (at the same time increasing in size from 169 to 306 pages with corresponding increases in detail and complexity).

2.4.3 Further reforms and the Building Act 1984

In 1974, the reorganisation of local government boundaries (brought about by the *Local Government Act 1972*) reduced the number of local authorities to about 400.

In many cases reorganisation created co-ordination and control problems for the new expanded local authorities and led to an increase in staff numbers and bureaucracy.

Although the building control system was seen to produce safe buildings in which fire and serious structural failure were rare, during the 1970s persistent criticisms existed on other counts. In particular, the system was perceived to be more cumbersome and bureaucratic than necessary. Additionally, the detailed form of the Regulations was blamed for inflexibility in use and was said to inhibit innovation and impose unnecessary costs.

In December 1979 a speech by the Secretary of State for the Environment launched a radical review of the building control system, in which he set out the criteria which any new arrangements would need to satisfy. These were:

- maximum self-regulation
- minimum Government interference
- total self-financing
- simplicity in operation.

The speech elicited a wide response, and in June 1980 the Department of the Environment issued a Consultation Document outlining a number of specific possibilities for change, drawn up in the light of the comments received. Amongst other items, the possibilities included:

- recasting the Building Regulations as a minimum number of functional requirements, with supporting approved Codes of Practice;
- 'certification' of plans and construction by approved private persons, as an optional alternative to local authority control.

The Consultation Document was circulated widely for comment and there was general agreement regarding the proposals for a revised form of Building Regulations. On the issue of private certification views were more deeply divided. A significant minority argued that control should remain solely with local authorities. However, the preponderant view, shared by some local authorities, was that certification would be a useful alternative to local authority checking, provided that the detailed arrangements for selecting certifiers and regulating their functions were sufficiently tightly drawn to safeguard public health and safety.

In February 1981 the Secretary of State for the Environment issued a command paper entitled '*The Future of Building Control in England and Wales*' (Cmnd 8179, 1981). This paper contained a range of proposals dealing with revision of the Building Regulations and with the processes of control. It included detailed proposals which would allow private 'certifiers' to check plans and inspect work on site. The criteria for approval by the Secretary of State would relate to the possession, by the private certifier, of professional qualifications, practical experience and indemnity insurance. Private certifiers could be individuals or corporate bodies.

On the basis of this document the Department of the Environment set about the task of completely recasting the Building Regulations and of producing a new

Act of Parliament which would amend the law relating to the supervision of work subject to Building Regulation control.

Regrettably, these requirements were coupled with some very controversial housing legislation and the intervention of a general election, and although the Housing and Building Control Bill was first placed before Parliament in 1982, the Housing and Building Control Act did not become law until June 1984. The Act contained the enabling legislation for:

- the supervision of building work otherwise than by local authorities (i.e. by Approved Inspectors)
- exemption of public bodies from the procedural requirements of building regulations
- the approval of documents for the purposes of giving guidance on building regulations
- approval of persons to give certificates of compliance with building regulations.

In the event, the parts of the *Housing and Building Control Act 1984* relating to building control were relatively short lived. They were subsumed into the *Building Act 1984*, which came into force on 1 December 1984. This Act consolidated various building control statutes enacted over the previous 90 years and helped to ease considerably the task of those engaged in building control. The *Building Act 1984* (as amended) is the principal controlling legislation for both the Local Authority and Approved Inspector building control processes, and is described fully in other parts of this book.

2.4.4 The Building Regulations 1985

Made under the *Building Act 1984*, the 1985 Regulations were the first to be cast wholly in functional terms. As they contained statements describing how the building and its services were required to function in relation to health, safety, welfare and convenience, the conservation of fuel and power, and the conservation of water, the regulations were short and concise. Unlike the 1976 Regulations, no technical details were given in the 1985 Regulations. For this (with the exception of means of escape in case of fire and facilities for disabled people) reference had to be made to a series of Approved Documents, compliance with which was not mandatory (again unlike the 1976 Regulations). Each of the following parts of the 1985 Regulations had equivalent Approved Documents:

- Part A: Structure
- Part B: Fire (excluding means of escape)
- Part C: Site preparation and resistance to moisture
- Part D: Toxic substances
- Part E: Resistance to the passage of sound
- Part F: Ventilation
- Part G: Hygiene

- Part H: Drainage and waste disposal
- Part J: Heat-producing appliances
- Part K: Stairways ramps and guards
- Part L: Conservation of fuel and power
- Regulation 7 (materials and workmanship).

Initially, means of escape in case of fire could only be dealt with by reference to a publication entitled '*The Building Regulations 1985 – Mandatory Rules for Means of Escape in Case of Fire*'. Additionally, the mandatory rules applied only to:

- a dwelling-house of three or more storeys (newly erected or created by material alteration or change of use)
- a flat in a building of three or more storeys
- an office
- a shop.

Compliance was achieved by reference to selected parts of four named British Standards. This had the effect of making the Standards mandatory documents.

Additionally, facilities for disabled people were dealt with by reference to Schedule 2 of the 1985 Regulations. This Schedule was identical to Part T of the 1976 Regulations (which had only been in existence for 3 months before the 1985 Regulations came into force) and applied to new shops, offices, single-storey factories, educational buildings and other new single-storey buildings open to the public. Unlike the technical requirements supporting the regulations concerning means of escape in case of fire which were mandatory, the old fashioned notion of '*deemed-to-satisfy provisions*' was used to provide technical guidance for facilities for disabled people. In this case reference was made to:

- *Design Note 18: Access for Disabled People to Educational Buildings* for schools and other educational establishments
- BS 5810: 1979 clauses 6.2 to 8.4.4 for the other buildings covered by Schedule 2.

If these documents were followed, compliance with the regulations could be guaranteed, but the applicant was not compelled to use them if compliance could be demonstrated in some other way.

The 1985 Regulations were amended twice before being replaced by the *Building Regulations 1991*. The most significant change to the 1985 Regulations was the introduction of a new Part M (with an accompanying Approved Document) covering access and facilities for disabled people, thereby replacing Schedule 2 of the 1985 Regulations.

2.4.5 The Building Regulations 1991

The *Building Regulations 1991* (the 1991 Regulations) came into force on 1 June 1991 thus replacing the 1985 Regulations and consolidating the previous amendments. The main change brought about by the 1991 Regulations was the abandonment of the *Mandatory rules for means of escape in case of fire* and their replacement in Schedule 1 with a new requirement B1 covering means of escape in case of fire together with an accompanying section in Approved Document B. The 1991 Regulations were amended eight times. The most significant changes were as follows:

- A new regulation 13A introducing regularisation certificates, which can be given to regulate unauthorised building work commenced after 11 November 1985.
- A new regulation 14A requiring energy rating for new dwellings.
- The revocation of regulation A4 dealing with disproportionate collapse.
- New requirements introduced in 1994 governing Part F (*Ventilation*) and Part L (*Conservation of fuel and power*).
- New requirements introduced in 1997 governing Part K (name changed to *Protection from falling, collision and impact*) including:
  - the extension of requirements K1 and K2 to apply to stairs, etc. which give access to levels used only for maintenance purposes
  - amendment of requirement K3 to protect people from colliding with vehicles in loading bays
  - new requirement K4 to protect people from collision with open windows, skylights and ventilators
  - new requirement K5 to protect people from being struck or trapped by sliding or powered doors or gates.
- New requirements introduced in 1997 governing Part N (name changed to *Glazing – materials and protection*) including:
  - new requirement N3 requiring the safe opening and closing, or adjusting of windows, skylights or ventilators
  - new requirement N4 requiring the provision of safe access for cleaning windows, skylights, or other transparent or translucent surfaces.
- New requirements introduced in 1999 governing Part M (*Access and facilities for disabled people*) extending the scope of Part M to dwellings.
- Revised regulation 7 (*Materials and workmanship*) introduced in 1999.
- Revisions to Part B (*Fire*) introduced in 2000 including a new Approved Document B. Significant changes to the guidance and the extension of requirement B1 to cover means of warning and escape.
- The introduction of a new Part IIA (*Exemption of public bodies from procedural requirements*) and a new regulation 9A (*The Metropolitan Police Authority*) exempting the Metropolitan Police Authority from the procedural requirements of the Building Regulations.

2.4.6 The Building Regulations 2000

The *Building Regulations 2000* (the 2000 Regulations) came into force on 1 January 2001, thus replacing the 1991 Regulations and consolidating the previous

amendments, their main purpose being to renumber certain regulations conse-
quent on the consolidation and to reword certain requirements to offer greater
clarity of interpretation. Since coming into force the 2000 Regulations have been
amended on six occasions (see Section 3.3.1). The main effects of the various
amendments are as follows:

2001 amendment (coming into force on 1 April 2002):

- Revisions to provisions for drainage in the *Building Act 1984*:
  – repeal of section 18 (building over sewers, etc.) – see Section 3.2.3;
  – repeal of subsections (1) and (2) of section 21 (provision of drainage) – see
    Section 3.2.3;
  – revision of section 59 (drainage of buildings) – so as to incorporate the sub-
    stance of subsection (2) of section 21.
- Revisions to provisions for drainage in Schedule 1 to the 2000 Regulations:
  – paragraphs H1 and H3 reworded so that the provision of foul and rainwater
    drainage is expressly required;
  – paragraph H2 retitled, *Wastewater treatment systems and cesspools*;
  – new paragraph H4 dealing with the building over of sewers (replacing section
    18 of the *Building Act 1984*);
  – new paragraph H5 dealing with the provision of separate systems of
    drainage;
  – the original paragraph H4, *Solid waste storage*, renumbered H6 and revised.
- Revised Part J of Schedule 1 to the 2000 Regulations retitled *Combustion
  appliances and fuel storage systems* including:
  – new paragraph J4 dealing with the provision of information;
  – new paragraph J5 covering the protection of liquid fuel storage systems;
  – new paragraph J6 dealing with protection against pollution.
- Revised Part L of Schedule 1 to the 2000 Regulations, *Conservation of fuel and
  power* including:
  – new paragraph L1 dealing exclusively with dwellings;
  – new paragraph L2 dealing exclusively with buildings other than dwellings.
- Revisions to the 2000 Regulations including:
  – new regulation 14A requiring the local authority to consult the sewerage
    undertaker where there are proposals to build over a sewer;
  – revised regulation 18, *Testing of building work*.

2002 amendment (coming into force on 1 April 2002):

- New regulation 16A, *Provisions applicable to replacement windows, rooflights,
  roof windows and doors* dealing with the system of self-certification for the
  installation of the listed items under the Fenestration Self-Assessment (FENSA)
  Scheme.
- New Schedule 2A, *Exemptions from requirement to give building notice or
  deposit full plans*.

2002 amendment No. 2 (coming into force between 1 March 2003 and 1 July 2004):

- New regulation 20A dealing with sound insulation testing.
- Revised regulation 8, *Limitation on requirements*.
- Revised Part B of Schedule 1 to the 2000 Regulations, *Fire safety*, including minor rewording of paragraph B2, *Internal fire spread (linings)*.
- Revised Part E of Schedule 1 to the 2000 Regulations, *Resistance to the passage of sound* including:
  - paragraph E1 retitled *Protection against sound from other parts of the building and adjoining buildings*
  - paragraph E2 retitled *Protection against sound within a dwelling-house etc.*
  - paragraph E3 retitled *Reverberation in common internal parts of buildings containing flats or rooms for residential purposes*
  - new paragraph E4 dealing with acoustic conditions in schools.

2003 amendment (coming into force on 1 December 2003 and 1 May 2004):

- Revised Part M of Schedule 1 to the 2000 Regulations retitled *Access to and use of buildings* including:
  - paragraph M1 retitled *Access and use*;
  - paragraph M2 retitled *Access to extensions to buildings other than dwellings*;
  - paragraph M3 retitled *Sanitary conveniences in extensions to buildings other than dwellings*;
  - paragraph M4 retitled *Sanitary conveniences in dwellings*.

2004 amendment (coming into force on 1 July 2004 and 1 December 2004):

- Revised Part C of Schedule 1 to the 2000 Regulations retitled *Site preparation and resistance to contaminants and moisture* including:
  - paragraph C1 retitled *Preparation of site and resistance to contaminants*
  - paragraph C2 retitled *Resistance to moisture*
  - paragraphs C3 and C4 removed.
- Revised Part F of Schedule 1 to the 2000 Regulations by the omission of paragraph F2, *Condensation in roofs*.

2004 amendment No. 2 (coming into force on 1 January 2005):

- Regulation 16A retitled *Provisions applicable to self-certification schemes* substantially revised to take account of a range of such schemes.
- New Part P of Schedule 1 to the 2000 Regulations entitled *Electrical safety* including the following:
  - paragraph P1, *Design, installation, inspection and testing*
  - paragraph P2, *Provision of information*.
- New Schedule 2B *Descriptions of work where no building notice or deposit of full plans required*.

2.4.7 The Approved Inspector system of control and the Building (Approved Inspectors etc) Regulations 1985 to 2000

It has taken some considerable time for the private certification system to become fully operational even though the first Approved Inspector, National House-Building Council (NHBC) Building Control Services Ltd (NHBC BCS Ltd), was approved on the 11 November 1985. Their original approval related only to dwellings of not more than four storeys (or three storeys and a basement) but this was later extended to include residential buildings up to eight storeys and this was further extended to include any buildings, in 1998.

The appointment of further Approved Inspectors was held up by a number of factors, but was mainly due to the difficulty posed in obtaining the level of insurance cover required by the Department of the Environment and agreement over such items as qualification and experience levels expected. After a period of consultation, new proposals for insurance requirements for Approved Inspectors wishing to handle the control of commercial building work were agreed, and these were implemented on 8 July 1996.

This resulted in the approval by the Secretary of State, of three further corporate bodies on 13 January 1997. Others (including in excess of 30 non-corporate Approved Inspectors) have continued to be approved since that date, but at present, NHBC BCS Ltd remain the only body insured to deal with speculative domestic construction (i.e. self-contained houses, flats and maisonettes built for sale to private individuals). In this context, the Department for Transport, Local Government and the Regions (DTLR) issued insurance guidelines on 23 October 2001 which allowed Approved Inspectors to carry out their building control function on a range of different dwelling types, except the so-called 'non-exempt' dwellings. This definition excludes speculative dwellings constructed by house-building companies for sale to the public.

On 8 July 1996 the Construction Industry Council (CIC) was designated as the body responsible for approving non-corporate Approved Inspectors, although initially, the Secretary of State reserved the right to approve corporate bodies. From 1 March 1999 the CIC became responsible also for the approval of corporate Approved Inspectors.

The Approved Inspector system of control is governed by sections 47 to 53 of the *Building Act 1984* and the *Building (Approved Inspectors etc) Regulations*. The first regulations (the 1985 Regulations) were made on 11 July 1985 and came into operation on 11 November 1985. The 1985 Regulations were amended five times mostly in response to changes in the principal regulations (i.e. the Building Regulations 1985, 1991 and 2000) before being repealed and replaced by the *Building (Approved Inspectors etc) Regulations 2000* (the 2000 Regulations) which came into force on 1 January 2001. The 2000 Regulations consolidated all the previous amendments and have since been amended four times in response to changes to the principal regulations.

Thus, over the last 20 years, building control by Approved Inspectors has become a viable alternative to the services offered by local authorities. Free competition exists for the control of building work in all but the speculative housing sector, and it is anticipated that new insurance proposals for this sector (which include proposals for a 'warranty link scheme' similar to that provided under the NHBC Buildmark scheme), currently being considered by Office of the Deputy Prime Minister (ODPM), will permit increased competition in this area of work in the near future.

2.5 Conclusions

This brief historical account has traced the building control system in England and Wales back to its origins in the new industrial world of the Victorian era and it has been shown that controls over construction were originally introduced in response to public health problems caused by overcrowding, poverty, ignorance and greed. Although public health was undoubtedly the catalyst which led to the foundation of the system it must be remembered that in the middle of the nineteenth century it was by no means clear that sub-standard housing and poor sanitation were the causes of the epidemics that regularly took such a toll of the population. It is also apparent that any new initiative which was likely to affect the wealth of the small numbers of the gentry who owned the vast proportion of the land was likely to meet substantial resistance in the very place where laws could be made; that is, Parliament. Therefore, the first attempts at control were greatly resisted until the political will for reform was sufficiently strong to override vested interests. So even in the early years, political considerations were paramount.

For over a hundred years public health and safety remained the key reason to make byelaws (and later, the regulations), and the principal means for ensuring compliance remained in the public sector with local authorities.

There is no doubt that the changes in the regulations and control systems which took place in the last quarter of the twentieth century were also politically motivated since concern for the economy (the oil crises of the 1970s) and the environment (global warming) led to controls over the conservation of fuel and power, and the production of greenhouse gases. Socio-political issues also became important with the introduction of regulations designed to ensure that buildings were convenient for use by disabled people.

However, the ultimate political intervention must be the influence of Thatcherite monetarism in the 1980s which led to the semi-privatisation of the building control system and the establishment of Approved Inspectors in competition with local authorities.

For the future, there is no doubt that the scope of issues covered by the regulations will increase. The most recent Consultation Document distributed by the ODPM concerns the desirability of installing ductwork, etc. capable of distributing electronic communication services (Broadband) around buildings.

PART 2

Legal background

3

Controlling legislation

3.1 Introduction

This chapter outlines the key legislative provisions which govern the control of building work in England and Wales. No attempt has been made to deal with the building control systems in Scotland and Northern Ireland, for although the regulations throughout the UK are all based on the same fundamental principles, there are subtle differences in form and content, and major differences in the way the regulations are administered between the different countries.

3.1.1 General structure of the building control system in England and Wales

The structure of building control in England and Wales is based on a system of primary and secondary legislation supported by non-mandatory technical guidance.

The principle source of primary legislation is the Building Act 1984 which received Royal Assent on 31 October 1984. Most of its provisions came into force on 1 December 1984. Those that did not are discussed in Section 3.2.2 below. The Act has been revised by other Acts of Parliament on numerous occasions, therefore a copy of the original Act of 1984 is really only of academic interest to the practitioner. An updated version of the complete Act can be found in *Knight's Guide to Building Control Law and Practice*.

The Building Act covers the making and administration of Building Regulations (both the Local Authority and Approved Inspector Systems) and also deals with a range of other provisions, such as drainage and the local authority's powers in relation to dangerous structures, defective premises, demolitions, etc. The Act is of prime importance and many of its provisions are explained in detail in other parts of this book.

Both the Building Regulations and the Building (Approved Inspectors etc) Regulations are made under powers contained in section 1 of the Building Act and constitute the principle pieces of secondary legislation in the overall structure.

The Building Regulations 1985 and the Building (Approved Inspectors etc) Regulations 1985 were the first to be made under the 1984 Act, and both came into force on 11 November 1985. There have been numerous revisions to both sets of regulations since 1985 and even the current 2000 editions of both sets of regulations have been revised several times.

The 1985 Building Regulations differed considerably from the previous regulations (the Building Regulations 1976) in that they were expressed in functional terms (unlike the 1976 Regulations which were mainly prescriptive). This made the 1985 Regulations simpler and easier to understand, and they were intended to be more flexible in use for the designer and/or builder. Conversely, the 1976 Regulations contained mandatory technical solutions, which tended to lead to inflexibility in interpretation, stifled innovation and, in many cases, could be more costly than was strictly necessary.

Since the new regulations were expressed in functional terms (i.e. they stated what had to be achieved without telling the designer how to achieve it) there was

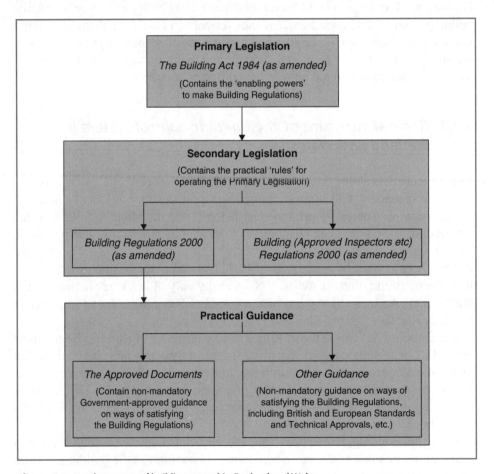

Figure 3.1 Legal structure of building control in England and Wales.

obviously a need to provide some form of guidance on how the requirements could be met. Initially, this was achieved by recasting many of the former technical solutions in the 1976 Regulations into the new 'Approved Documents'. However, in the process the technical solutions lost their mandatory status and became merely recommendations and guidance, albeit approved by the Government for the purpose of satisfying the Building Regulations.

Thus the legal structure of the current building control system in England and Wales can be summarised in Figure 3.1.

Other regulations made under the powers contained in the Building Act include the *Building (Inner London) Regulations 1985 and 1987* (see Section 4.3.10) and the *Building (Local Authority Charges) Regulations 1998* (see Section 4.3.7).

3.2 The Building Act 1984

3.2.1 Introduction

The Building Act 1984 received the Royal Assent on 31 October 1984 and the majority of its provisions came into force on 1 December 1984. The Act was primarily a consolidating measure and made no major changes in the law. It did, however, incorporate Parts II and III of the *Housing and Building Control Act 1984* which had not, in fact, been put into effect. Since these parts dealt with the form and structure of the now current building control system, this incorporation was of paramount importance to the future of building control in England and Wales. The Building Act also brought together a number of provisions relating to buildings, which had previously been contained in a number of other statutes including:

- The Public Health Acts 1875, 1936 and 1961
- The Water Acts 1945, 1973 and 1981
- The Offices, Shops and Railway Premises Act 1963
- The Fire Precautions Act 1971
- The Control of Pollution Act 1974
- The Local Government (Miscellaneous Provisions) Acts 1976 and 1982
- The Health and Safety at Work etc Act 1974
- The Local Government Acts 1963 and 1972
- The Education Acts 1944 and 1980
- The London Government Act 1963.

This consolidation was of considerable assistance to building practitioners since it provided most of the necessary legislation in one Act and avoided the need to search through a confusing array of enactments concerning buildings and related matters.

3.2.2 Commencement of the provisions of the Building Act 1984

Whilst most sections of the Building Act 1984 came into force on 1 December 1984 some sections have yet to be activated. Section 134 of the Act sets out details of those provisions, which are currently in force and gives details of those sections which may be brought into effect at a later date. At present the following sections have not yet been activated:

- Sections 12, 13, 31, 38, 42(4) to 42(6) and 43(3) (except so far as they enable regulations to be made).
- Sections 20, 33, 42(1) to 42(3), 43(1) and 43(2), 44 and 45.
- Section 132(2) and Schedule 7 so far as they relate to the Atomic Energy Act 1954.

Sections 50(2) and 50(3) were activated on 11 November 1985 (by virtue of the *Building Act 1984 (Commencement No. 1) Order 1985*) and paragraph 9 of Schedule 1 came into force on 7 August 1998 (by virtue of the *Building Act 1984 (Commencement No. 2) Order 1998*). Paragraph 9 enabled the Secretary of State to make regulations authorising local authorities to fix and recover charges for and in connection with the performance of their functions under the building regulations. This resulted in the coming into force of the *Building (Local Authority Charges) Regulations 1998* (see Section 4.3.7).

Some of the above sections which have yet to come into force are of particular interest as follows:

- **Section 20** deals with the use of materials which are unsuitable for permanent building and is designed to supersede section 19 (use of short-lived materials) on a day to be appointed by the Secretary of State. Section 20 is wider in scope than section 19 since it refers not only to materials and components unsuitable for permanent building but also to prescribed services, fittings or items of equipment.
- **Section 33** covers tests for conformity with building regulations and enables a local authority to require any person who is carrying out building work (or on whose behalf it is being carried out) to carry out such tests as may be specified, to ascertain if the building regulations would be contravened. This provision may be contrasted with requirement regarding testing in the Building Regulations (see Appendix 1, regulation 18), whereby it is up to the local authority to make such tests of any building work as may be necessary to establish whether it complies with regulation 7 or any of the applicable requirements contained in Schedule 1. Section 33 also makes it clear that the cost of testing is to be met by the person carrying out the test. Regulation 18 does not comment on this aspect and it is a source of confusion especially where the testing can be expensive (such as airtightness testing of building under Parts L1 and L2, see Sections 5.4.9 and 5.8.2).

- **Section 44** covers the application of the building regulations to the Crown (see also Section 4.2.1). It provides for the substantive provisions of the regulations to be applied to work carried out by or on behalf of the Crown authority, whether or not in relation to a Crown building.

3.2.3 Repealed sections of the Building Act 1984

Whilst there have been a great many revisions to the 1984 Act since it first came into force, a number of sections have been entirely repealed or substantially curtailed by later legislation. Often this has been as a result of the decision to expand the provisions of the building regulations to cover some sections of the Building Act. This has the advantage of allowing guidance to be given (via the Approved Documents) which reduces the vagaries of interpretation, which had previously occurred between different local authorities. The following sections are of particular interest:

- **Section 18** *Building over a sewer*. This section was repealed by regulation 3 of the *Building (Amendment) Regulations 2001* which came into force on 1 April 2002. Its provisions were replaced by a new requirement H4 in Part H of Schedule 1 to the *Building Regulations 2000*. One result of the introduction of H4 was to take away from local authorities the exclusive rights to administer provisions regarding the building over of a sewer since this can now also be exercised by the Approved Inspectors.
- **Section 21** *Provision of drainage*. Paragraphs (1) and (2) of this section (which allowed a local authority to reject plans of a proposal which did not show adequate provisions for drainage of the building) were repealed by regulation 3 of the *Building (Amendment) Regulations 2001* which came into force on 1 April 2002. This removed a curious anomaly in the system whereby, although the building regulations contained requirements regarding the drainage of buildings they could not require buildings to be drained! This was the purpose of section 21. Part H of Schedule 1 has now been extended to require buildings to be adequately drained, and where the local authority is used for the purposes of building control, any full plans deposited with the local authority must be accompanied by particulars of the provision to be made for drainage. In cases where building control is supplied by an Approved Inspector it is not necessary to provide the Local Authority with the design of the drainage system. However it will be necessary to provide with the initial notice an indication of the location of any connection to be made to a sewer, or the proposals for the discharge of any proposed drain or private sewer including the location of any septic tank and associated secondary treatment system, or any wastewater system or any cesspool. Those parts of section 21, which have not been repealed are discussed in Section 3.2.4.
- **Section 23** *Provision of facilities for refuse*. Paragraphs (1) and (2) of this section (which allowed a local authority to reject plans of a proposal which did not

show adequate provisions for the storage and removal of refuse) were repealed by regulations 18 and 20 of the *Building Regulations 1985*. Requirement H6 of Part H of Schedule 1 to the Building Regulations 2000 now covers this provision. Those parts of section 23, which have not been repealed are discussed in Section 3.2.4 below.

- The following sections of the 1984 Act have all been completely repealed:
  - Section 26 *Provision of closets*, replaced by requirement H1 in Part H of Schedule 1 to the Building Regulations
  - Section 27 *Provision of bathrooms*, replaced by requirement G1 in Part G of Schedule 1 to the Building Regulations
  - Section 28 *Provision of food storage*, not replaced
  - Section 29 *Site containing offensive material*, replaced by requirement C1 in Part C of Schedule 1 to the Building Regulations
  - Section 30 *Determination of questions*, not replaced
  - Section 69 *Provision of water supply in occupied house*, repealed by the *Water Act 1989* (various sections).

3.2.4 Layout, structure and analysis of the Building Act 1984

The Act is laid out in five parts and contains seven Schedules. Where detailed descriptions of the sections have been given in other parts of this book, this is indicated against the relevant section.

Building Act 1984: Part I

Part I of the 1984 Act deals with Building Regulations and is structured as follows.

Power to make building regulations

Section 1   Power to make building regulations

From the perspective of building regulations in England and Wales, this section is the most important in the 1984 Act, for without it the system would not exist. This is what is known as an 'enabling power' since it enables the Secretary of State to make the regulations. The full text of section 1 is as follows:

(1) The Secretary of State may, for any of the purposes of:
  (a) securing the health, safety, welfare and convenience of persons in or about buildings and of others who may be affected by buildings or matters connected with buildings
  (b) furthering the conservation of fuel and power
  (c) preventing waste, undue consumption, misuse or contamination of water
  make regulations with respect to the design and construction of buildings, and the provision of services, fittings and equipment in or in connection with buildings.

(2) Regulations made under Subsection (1) above are known as building regulations.

(3) Schedule 1 to this Act has effect with respect to the matters as to which building regulations may provide.

(4) The power to make building regulations is exercisable by statutory instrument which is subject to annulment in pursuance of a resolution of either House of Parliament.

Subsection (1) lays down the purposes for which building regulations may be made. It should be noted that this does not extend to the protection of property. Subsection (3) refers to Schedule 1 of the 1984 Act where the matters which may be provided for in building regulations are set out in some detail (see below). These matters should not, however, be seen to limit the generality of subsection (1). Building regulations are known as a 'statutory instrument' since they are a means of bringing into force the generalised provisions of the Act (i.e. the 'statute'). Therefore they are not subject to the full Parliamentary scrutiny afforded to an Act of Parliament. The procedure for making such a statutory instrument is known as the 'negative procedure' and is laid down in section 5 of the *Statutory Instruments Act 1946*. Statutory instruments do not go through the procedure of Parliamentary readings and debate; instead they are merely presented to Parliament and will become law at the end of a prescribed period of time unless a resolution is passed by either House of Parliament for annulment (see also Section 7.1).

Section 2   Continuing requirements

Under this section building regulations can be made to impose continuing requirements on owners and occupiers of buildings. There are two kinds of continuing requirements:

- those in respect of a designated provision of the Building Regulations to ensure that the purpose of the provision is not frustrated
- those with regard to services, fittings and equipment.

This enables requirements to be imposed on buildings whenever they were erected and independently of the normal application of building regulations to the building.

For example, where an item is required to be provided, such as mechanical ventilation designed to reduce condensation in a building, there could be a requirement that it should continue to be provided or kept in working order. The power could also be used to require the maintenance and periodic inspection of equipment, such as boilers and air conditioning plant. No regulations have yet been made under this section.

Exemption from building regulations

Section 3   Exemption of particular classes of buildings etc.

Building regulations may exempt a prescribed class of buildings (including a particular building, or buildings of a particular class at a particular location), services, fittings or equipment from all or any of the provisions of the Building Regulations.

This can be done with or without conditions. To date the powers in section 3 have been used to exempt the seven classes of buildings described in Schedule 2 to the *Building Regulations 2000* (see Section 4.2.4) and certain categories of work from the need to pay charges under the *Building (Local Authority Charges) Regulations 1998* (see Section 4.3.7).

Section 4   Exemption of educational buildings and buildings of statutory undertakers (see Sections 4.2.2 and 4.2.3)

Section 5   Exemption of public bodies from procedural requirements of building regulations

Under this section exemption may be granted to a local authority, a county council and any other prescribed non-profit-making public body, from the need to comply with the *procedural* requirements (i.e. the administrative provisions) of building regulations. Such exempt bodies must still comply with the substantive requirements (i.e. the technical provisions) of the regulations but are immune from prosecution by the local authority under section 35 of the 1984 Act (and therefore, from having to remove or alter offending work under section 36). To date, only one public body (the Metropolitan Police Authority) has been exempted under section 5 (see Appendix 1, Part III, regulation 10).

Approved documents

Section 6   Approval of documents for purposes of building regulations (see Sections 7.2 and 7.3)

Section 7   Compliance or non-compliance with approved documents (see Sections 7.2 and 7.3)

Relaxation of building regulations

Section 8   Relaxation of building regulations (see Section 6.3.3)

Section 9   Application for relaxation (see Section 6.3.3)

Section 10   Advertisement of proposal for relaxation of building regulations (see Section 6.3.3)

Section 11   Type relaxation of building regulations

The Secretary of State may make a relaxation or dispensation which applies generally to a particular type of building matter if he feels that the operation of the building regulations would be unreasonable when applied to that building matter. The type relaxation may be made with or without conditions, but the Secretary of State is required to consult interested parties before giving the direction. Although a number of type relaxations were given under earlier legislation, none have been issued with respect to any building regulations made under the 1984 Act.

Type approval of building matter

Section 12   Power of Secretary of State to approve a type of building matter

Under this section the Secretary of State may approve types of building matter as complying either generally or with particular requirements of building regulations.

This is to be done by the issue of a certificate to that effect. This power can be delegated to any person or body by the Secretary of State under section 13 below. Both sections 12 and 13 have yet to be activated.

Section 13 Delegation of power to approve
See Section 12 above.

Consultation

Section 14 Consultation with the Building Regulations Advisory Committee and other bodies (see Section 7.2.1)

Section 15 Consultations with the Fire Authority
Before exercising their powers to dispense with or relax building regulations concerning structural fire precautions or means of escape in case of fire, the Local Authority must consult the Fire Authority. See also Sections 5.3.2 and 5.7.1.

Passing of plans

Section 16 Passing or rejection of plans (see Section 5.2.2)

Section 17 Approval of persons to give certificates
The Secretary of State is empowered by this section to provide in building regulations for the approval of persons to certify plans for the purpose of section 16(9) of the 1984 Act (see Section 5.2.2). This provision has been made in regulation 29 of the *Building (Approved Inspectors etc) Regulations 2000*. The procedure for approval of persons is similar to that for the approval of Approved Inspectors (see Section 5.5).

The following bodies, together with the Chartered Institution of Building Services Engineers, have been designated to approve private individuals who wish to become approved persons who can certify plans to be deposited with the local authority as complying with the energy conservation requirements:

- The Chartered Institute of Constructors
- The Faculty of Architects and Surveyors
- The Association of Building Engineers
- The Institution of Building Control Officers
- The Institution of Civil Engineers
- The Institution of Structural Engineers
- The Royal Institute of British Architects
- The Royal Institution of Chartered Surveyors.

Additionally, the Institution of Civil Engineers and the Institution of Structural Engineers have been designated to approve persons to certify plans as complying with the structural requirements. As yet however, no approved persons have been designated in England and Wales although a pilot scheme for structural approvals has been operating in Scotland for some time with limited success.

Section 18 Building over sewers etc. (see Sections 3.2.3, 5.3.1 and 5.7.2)

Section 19 Use of short-lived materials

The use of materials which are unsuitable for permanent buildings is covered by section 19 of the 1984 Act. Local authorities are enabled to reject plans for the construction of buildings of short-lived or otherwise unsuitable materials, or to impose a limit on their period of use. The Secretary of State may, by building regulations, prescribe materials which are considered unfit for particular purposes. Tables 1 and 2 of the 1976 Regulations listed materials, which were considered unfit for the weather-resisting part of any external wall or roof. Neither the 2000 regulations nor Approved Document to support regulation 7 prescribe any materials as unfit for particular purposes as yet, however the Approved Document does lay down some general criteria against which materials may be judged. Bearing this in mind, it is unlikely that section 19 can be used by local authorities to proscribe certain materials.

Section 20   Use of materials unsuitable for permanent building (see Section 3.2.2)

Section 21   Provision of drainage

Although sub-paragraphs (1) and (2) of this section have been replaced by requirement H1 of Schedule 1 to the Building Regulations 2000 (see this Section 3.2.3), a local authority (or on appeal a magistrates' court) may still require a proposed drain to connect with a sewer where that sewer is within 100 feet of the site of the building. In cases where the sewer is located more than 100 feet from the site of the building, the local authority may still require connection to that sewer if they undertake to bear the additional cost (i.e. for the length of drain in excess of 100 feet) of construction, maintenance and repair. Disputes regarding the cost of the additional work may be referred to the magistrates' court.

Section 22   Drainage of building in combination

The powers of a local authority under section 21 are extended by this section so that where two or more buildings are involved, the local authority may require them to be drained in combination (instead of each making a separate connection) into an existing sewer. As for section 21 the drain may be constructed by the owners (or by the local authority on their behalf) and the expenses of construction, maintenance and repair may be proportioned between each owner and the local authority as appropriate. Disputes regarding the cost of the apportionment may be referred to the magistrates' court.

Section 23   Provision of facilities for refuse

Although sub-paragraphs (1) and (2) of this section have been replaced by requirement H6 of Schedule 1 to the Building Regulations 2000 (see this Section 3.2.3), under subsection (3) of section 23 it is an offence for any person to close or obstruct a means of access for removal of refuse without the consent of the local authority. This may be granted with conditions.

Section 24   Provision of exits etc.

This section deals with the necessity to provide:

'such means of ingress and egress, and passages or gangways as the authority, after consultation with the fire authority, deem satisfactory, regard being had to the purposes for which the building is intended to be, or is, used and the number of persons likely to resort to it at any one time.'

Clearly, the intention of section 24 is that the relevant buildings should be provided with adequate means of escape.

Not all buildings are covered by this requirement since it refers only to:

- a theatre, and a hall or other building used as a place of public resort
- a restaurant, shop, store or warehouse which employs more than 20 people and to which members of the public are admitted
- a club licensed to serve intoxicating liquor (i.e. registered under the Licensing Act 1964)
- schools which are not exempt from the Building Regulations
- a church, chapel or other place of public worship, but this does not include a private house to which members of the public might be admitted only occasionally. Also excluded from control are any churches and chapels which were so used before 1 October 1937 (i.e. the date of commencement of the Public Health Act 1936).

Section 24 has its origins in pre-war public health legislation and has, to all intents and purposes, been superseded by requirement B1 (means of warning and escape) of Schedule 1 to the Building Regulations 2000, since it does not apply to any building or extension which is covered by B1. Furthermore, any local Acts of Parliament (see Section 4.3.2), which impose similar requirements to Section 24, are also superseded by B1.

Section 25   Provision of water supply (see Section 5.2.2)
Section 26   Provision of closets (see Section 3.2.3)
Section 27   Provision of bathrooms (see Section 3.2.3)
Section 28   Provision of food storage (see Section 3.2.3)
Section 29   Site containing offensive material (see Section 3.2.3)

Determination of questions
Section 30   Determination of questions (see Section 3.2.3)

Proposed departure from plans
Section 31   Proposed departure from plans (see Section 3.2.2)

Lapse of deposit of plans
Section 32   Lapse of deposit of plans (see Section 5.2.2)

Tests for conformity with building regulations
Section 33   Tests for conformity with building regulations (see this Section 3.2.2)

Classification of buildings
Section 34   Classification of buildings
For the purposes of building regulations, buildings may be classified under this section by reference to size, description, design, purpose, location or any other

characteristic whatsoever. The term 'building' is defined in section 121 of the 1984 Act (see Section 4.1.1) and in regulation 2(1) (see Appendix 1).

Breach of building regulations

Section 35   Penalty for contravening building regulations

A person contravening any provision in the building regulations (except for a provision which is designated in the regulations as one to which section 35 does not apply) is liable on summary conviction to a fine not exceeding level 5 on the 'standard scale' and to a further fine not exceeding £50 per day for a continuing offence (see Section 6.1.4, and section 114 of the 1984 Act below). The standard scale of fines for summary offences was introduced by section 37 of the *Criminal Justice Act 1982* (as amended by section 17 of the *Criminal Justice Act 1991*). The current levels of fines are as follows:

Level on standard scale	Amount of fine
Level 1	£200
Level 2	£500
Level 3	£1000
Level 4	£2500
Level 5	£5000

Regulation 17 of the *Building Regulations 2000* is a designated exclusion as are all of the *Building (Approved Inspectors etc) Regulations 2000* (except regulations 12 and 20).

Section 36   Removal or alteration of offending work (see Section 6.1.4)

Section 37   Obtaining a report where section 36 notice given (see Section 6.1.5)

Section 38   Civil liability (see Section 3.2.2)

Appeals in certain cases

Section 39   Appeal against refusal etc to relax building regulations (see Section 6.3.3)

Section 40   Appeal against section 36 notice (see Section 6.1.5)

Section 41   Appeal to Crown Court

An 'aggrieved person' against a decision of the magistrates under Parts I and IV of the 1984 Act may appeal to the Crown Court, unless each of the parties is entitled to require a dispute to be referred to arbitration (see section 106 of the 1984 Act below).

Section 42   Appeal and statement of case to High Court in certain cases (see Section 3.2.2)

Section 43   Procedure on appeal to Secretary of State on certain matters (see Section 3.2.2)

Application of building regulations to Crown etc
Section 44 Application to Crown (see Sections 3.2.2 and 4.2.1)
Section 45 Application to the UK Atomic Energy Authority (see Sections 3.2.2 and 4.2.3)

Inner London
Section 46 Inner London (see Section 4.3.10)

Part II of the 1984 Act deals with supervision of building work etc otherwise than by local authorities and is structured as follows.

Supervision of plans and work by Approved Inspectors
Section 47   Giving and acceptance of initial notice (see Section 5.6.5)
Section 48   Effect of initial notice (see Section 5.6.5)
Section 49   Approved Inspectors (see Section 5.5)
Section 50   Plans certificates (see Section 5.6.7)
Section 51   Final certificates (see Section 5.8.4)
Section 51A Variation of work to which initial notice relates (see Section 5.6.9)
Section 51B Effect of amendment notice (see Section 5.6.9)
Section 51C Change of person intending to carry out work (see Section 5.6.5)
Section 52   Cancellation of initial notice (see Sections 5.6.5 and 6.4)
Section 53   Effect of initial notice ceasing to be in force (see Section 5.8.4)

Supervision of their own work by public bodies
Section 54   Giving, acceptance and effect of public body's notice
This section, when read in conjunction with section 5 of the 1984 Act (see above) and Part VII of the *Building (Approved Inspectors etc) Regulations 2000* is concerned with public bodies and the detailed rules which allow such bodies to self-certify their own work. Public bodies are approved by the Secretary of State under section 5 of the 1984 Act. The regulations relating to notices, consultation with the fire authority, plans certificates and final certificates mirror those of Part III of the *Building (Approved Inspectors etc) Regulations 2000* dealing with Approved Inspectors (see Sections 5.6 and 5.9 *Building Control by Public Bodies*).

Supplementary
Section 55   Appeals
An 'aggrieved person' may appeal to a magistrates' court against the rejection by a local authority of any of the following:

- an initial notice, amendment notice or a public body's notice, or
- a plans certificate, final certificate, public body's plans certificate or a public body's final certificate.

There is a further right of appeal against the decision of the magistrates' court to the Crown Court.

Section 56   Recording and furnishing of information

Local authorities must keep a register, which is available for inspection by the public at all reasonable times, giving information about notices, certificates and insurance cover. See regulation 30 of the *Building (Approved Inspectors etc) Regulations 2000* in Section 3.4 below.

The information must be entered into the register as soon as practicable and not more than 14 days after the event to which it relates. Information about public body's notices and initial notices need only be kept on the register for as long as they are in force.

Section 57   Offences

If under Part II of the 1984 Act, a person gives a certificate or notice containing a statement, which he knows to be false or misleading in a material particular, he commits an offence under section 57. The same applies where he recklessly gives a notice or certificate containing a false or misleading statement. If an approved inspector or a person approved under section 17 of the 1984 Act, is convicted of an offence under this section, in addition to be fined or possibly sent to prison, his approval may be withdrawn immediately on receipt of the certificate of conviction forwarded by the court to the person by whom the approval was given. Such people may not be re-approved for at least 5 years.

Section 58   Construction of Part II

This section clarifies certain terms used in section II of the 1984 Act.

Part III of the 1984 Act deals with a range of other provisions about buildings. In many cases they give the local authority powers to control conditions and standards in existing buildings where the Building Regulations do not apply. These provisions can only be dealt with by Local Authorities (i.e. they cannot be dealt with by Approved Inspectors). Whilst many of these provisions may be dealt with by the local authority building control section, they are not directly connected with the Building Regulations so are not covered in detail in this book. They may be summarised as follows:

Sections 59–63	*Drainage*
Sections 64–68	*Provision of sanitary conveniences*
Sections 69–75	*Buildings* (for section 72 see Section 4.3.1)
Sections 76–83	*Defective premises, demolition etc* (see Section 4.3.1)
Sections 84 and 85	*Yards and passages*
Section 86	*Appeal to a Crown Court*
Section 87	*Application of provisions to Crown Property*
Section 88	*Inner London*
Sections 89 and 90	*Miscellaneous*

Section 90 is of interest in that it deals with facilities for inspecting local Acts of Parliament (see also Section 4.3.2). Where such an Act is in force containing provisions that impose an obligation or restriction as to the construction, nature or situation of buildings the local authority must keep a copy of those provisions at their offices for inspection by the public at all reasonable times free of charge.

Part IV of the 1984 Act is entitled 'General'. It brings together a number of disparate provisions only a few of which have direct relevance to building control. These are discussed in detail below.

Duties of local authorities

Section 91 Duties of local authorities

Local authorities are obliged to carry the 1984 Act into execution in their areas and must enforce building regulations (see Section 6.1.1).

Documents

Section 92 Form of documents

All notices given to or by a local authority must be in writing. Under decided case law, the term 'writing' may include reference to other modes of reproducing or representing words in visible form (see the *Interpretation Act 1978, section 5, Schedule 1*).

Section 93 Authentication of documents

This section describes those people in the local authority who are authorised to sign notices, orders, demands or other documents required by the 1984 Act. Any document bearing a signature of an officer who has not been authorised by the local authority will be invalid. However, it is possible for a document to be signed *per procurationem* (i.e. pp) by another officer if that person has been given written or oral authority by the properly authorised officer. The term 'signature' includes a facsimile of a signature by whatever process produced.

Section 94 Service of documents

Detailed provisions are given in this section for the service of notices and other documents under the 1984 Act in cases where no other provisions have been made.

Entry on premises

Section 95 Power to enter premises (see Section 6.1.2)

Section 96 Supplementary provisions as to entry (Section 6.1.2)

Execution of works

Section 97 Power to execute work

Certain sections of the 1984 Act permit local authorities to require the execution of works (e.g. see section 36 described in Section 6.1.4). Section 97 allows the local authority to carry out the work at the expense of the owner or occupier of the premises but it must be by agreement with that person. This also applies to any work carried out in connection with the construction, laying, alteration or repair of a sewer or drain provided that the owner or occupier is entitled to execute that work.

Section 98 Power to require occupier to permit work

This section operates to prevent an occupier of premises from frustrating the carrying out of work by an owner of premises in cases where the owner is required by the 1984 Act to execute that work. A complaint must be made by the owner to a magistrates' court. The court may then order the occupier to allow the execution of the work.

Of the remaining sections in Part IV of the 1984 Act very few of them have a direct bearing on building control or building regulations. Those that do are listed below.

Section 103 Procedure on appeal or application to a magistrates' court
Some sections of the 1984 Act allow an aggrieved person to appeal to a magistrates' court (e.g. see section 40 of the 1984 Act described in Section 6.1.5). Appeals must be brought within 21 days of the day on which the local authority's requirement, refusal or other decision was served on the person desiring the appeal. The right of appeal and the time within which it is to be brought must be stated in the local authority's decision document or it may be declared invalid. It should be noted that under sections 41 and 86 of the 1984 Act there is a further right of appeal to the Crown Court against the decision of a magistrates' court.

Section 104 Local authority to give effect to appeal
Where an appeal to a court varies or reverses the decision of a local authority it must give effect to the order of the court. This could include the grant or issue of any consent, certificate or other document and the making of an entry in any register.

Section 106 Compensation for damage
The local authority must pay full compensation to a person who has sustained damage as a result of the exercise of the local authority's powers provided that the person has not been in default. Interestingly, any dispute arising under this section as to the fact of damage, or as to the amount of compensation, must be determined by arbitration unless the compensation claim is less than £50.

Section 107 Recovery of expenses etc
Certain sections of the 1984 Act allow a local authority to charge expenses for the execution of work etc. (e.g. see section 97 above). Section 107 allows the local authority to recover those expenses at the time the works are completed. Reasonable interest may be charged from the date of service of the demand. If the ownership of the premises changes during the carrying out of the works or before the demand is served then the sums due are a charge on the premises and all estates and interests therein, and may be recovered from any person who subsequently acquires the property. Section 108 permits a local authority to allow the payment of the charges (with interest) by instalments over not more than 30 years.

Section 112 Obstruction
It is an offence for a person wilfully to obstruct a person acting in the execution of the 1984 Act, Building Regulations, or orders or warrants issued under the 1984 Act. 'Wilful' means intentional, deliberate or without lawful excuse but not merely unintentional. Obstruction does not need to be physical or involve violence, but could involve anything that makes it more difficult for a person to carry out his duty.

Section 114 Continuing offences
Some sections of the 1984 Act allow for fines to be imposed for a continuing offence (see section 19 *Use of short-lived materials* above). In the case of a

continuing offence the 1984 Act lays down certain fines which are payable for each day that the offence continues after a certain date. In most cases this will run from the expiry of any time fixed by the court for compliance with any directions given.

Interpretation

Certain sections within Part IV of the 1984 Act contain definitions of terms used throughout the Act. The most important of these for the purposes of building control are listed below.

Section 121 Meaning of 'building' (see Chapter 4 Section 4.1.1)
Section 122 Meaning of 'building regulations'
Any regulations made under section 1 of the 1984 Act (see section 1 above)
Section 123 Meaning of 'construct' and 'erect' (see Section 4.1.2)
Section 124 Meaning of deposit of plans
The deposit of plans under section 16 of the 1984 Act (see above and Section 5.2.2).

3.3 The Building Regulations 2000

3.3.1 Introduction

The *Building Regulations 2000 (Statutory Instrument 2000 No. 2531)* came into force on 1 January 2001. The 2000 Regulations revoked and replaced with amendments the *Building Regulations 1991* and consolidated all eight subsequent amendments to those regulations. Since coming into force the 2000 Regulations have been amended on six occasions by the following Statutory Instruments (SI):

- the Building (Amendment) Regulations 2001 (SI 2001 No. 3335)
- the Building (Amendment) Regulations 2002 (SI 2002 No. 440)
- the Building (Amendment) (No. 2) Regulations 2002 (SI 2002 No. 2871)
- the Building (Amendment) Regulations 2003 (SI 2003 No. 2692)
- the Building (Amendment) Regulations 2004 (SI 2004 No. 1465)
- the Building (Amendment) (No. 2) Regulations 2004 (SI 2004 No. 1808).

The 2000 Regulations are reproduced in full in Appendix 1 together with a commentary. This has been updated to include all the revisions brought about by the amendments referred to above. Where detailed descriptions of the regulations have been given in other parts of this book this is indicated against the relevant regulation.

This section should be read in conjunction with Section 3.2 and Appendix 1.

Regulations 1 to 24 in Parts I to VI are commonly referred to as 'procedural requirements' in the Building Regulations.

3.3.2 *Layout, structure and analysis of the Building Regulations 2000*

The Regulations are divided into six parts and there are four schedules as follows:

Part I   **General**
Regulation 1   Citation and commencement (see commentary to regulation 1 in Appendix 1)

Regulation 2   Interpretation (see commentary to regulation 2 in Appendix 1)

Part II   **Control of building work**
Regulation 3   Meaning of building work (see Section 4.1)
Regulation 4   Requirements relating to building work (see Section 4.1)
Regulation 5   Meaning of material change of use (see Section 4.1.6)
Regulation 6   Requirements relating to material change of use (see Section 4.1.7)
Regulation 7   Materials and workmanship (see Sections 4.3.8 and 7.5)
Regulation 8   Limitation on requirements (see Section 7.3.2)
Regulation 9   Exempt buildings and work (see Section 4.2.4)

Part III   **Exemption of public bodies from procedural requirements**
Regulation 10   The Metropolitan Police Authority (see Sections 4.2 and 5.9)

Part IV   **Relaxation of requirements**
Regulation 11   Power to dispense with or relax requirements (see Section 6.3.3)

Part V   **Notices and plans**
Regulation 12   Giving of a building notice or deposit of plans (see Section 5.2)
Regulation 13   Particulars and plans where a building notice is given (see Section 5.2)
Regulation 14   Full plans (see Section 5.2)
Regulation 14A   Consultation with sewerage undertaker (see Section 5.3)
Regulation 15   Notice of commencement and completion of certain stages of work (Section 5.4)
Regulation 16   Energy rating (see Sections 5.4.8 and 5.6.6)
Regulation 16A   Provisions applicable to self-certification schemes (see Section 5.11)
Regulation 17   Completion certificates (see Section 5.4.7)

Part VI   **Miscellaneous**
Regulation 18   Testing of building work (see Section 5.4.9)
Regulation 19   Sampling of material (see Section 5.4.9)
Regulation 20   Supervision of building work otherwise than by local authorities (see commentary to regulation 20 in Appendix 1)
Regulation 20A   Sound insulation testing (see Section 5.4.9)
Regulation 21   Unauthorised building work (see Section 6.2)
Regulation 22   Contravention of certain regulations not to be an offence (see Section 3.2.4 and commentary to regulation 20 in Appendix 1)

Regulation 23   Transitional provisions (see commentary to regulation 1(1) in Appendix 1)

Regulation 24   Revocations

The 2000 Regulations revoke the regulations listed in Schedule 3. This includes the *Building Regulations 1991* and eight subsequent amending regulations.

Schedules

Schedule 1   Requirements

Schedule 1 lists what are known as the 'substantive' requirements of the Building Regulations. These are reproduced in full with accompanying commentary, in Appendix 1. The substantive requirements are presented in a series of alphanumeric 'Parts' from Parts A to P (missing out the letters I and O) (see commentary in Appendix 1)

Schedule 2   Exempt Buildings and Work (see Section 4.2.4)

Schedule 2A   Exemptions from Requirement to Give Building Notice or Deposit Full Plans (see Section 5.11)

Schedule 2B   Descriptions of work where no building notice or deposit of full plans required (see Section 5.11.6)

Schedule 3   Revocation of Regulations (see Appendix 1)

3.4 The Building (Approved Inspector etc) Regulations 2000

3.4.1 Introduction

The *Building (Approved Inspectors etc) Regulations 2000 (Statutory Instrument 2000 No. 2532)* came into force on 1 January 2001. The 2000 Regulations revoked and replaced with amendments the *Building Regulations 1985* and consolidated all four subsequent amendments to those regulations. Since coming into force the 2000 Regulations have been amended on six occasions by the following Statutory Instruments:

- the *Building (Approved Inspectors etc) (Amendment) Regulations 2001 (SI 2001 No. 3336)*
- the *Building (Approved Inspectors etc) (Amendment) Regulations 2002 (SI 2002 No. 2872)*
- the *Building (Approved Inspectors etc) (Amendment) Regulations 2003 (SI 2003 No. 3133)*
- the *Building (Approved Inspectors etc) (Amendment) Regulations 2004 (SI 2004 No. 1466)*

The 2000 Regulations are reproduced in full in Appendix 2 together with a commentary. This has been updated to include all the revisions brought about by the amendments referred to above. Where detailed descriptions of the regulations have been given in other parts of this book this is indicated against the relevant regulation.

This section should be read in conjunction with Section 3.2 and Appendix 2.

3.4.2 Layout, structure and analysis of the Building (Approved Inspectors etc) Regulations 2000

The Regulations are divided into eleven parts and there are eight schedules as follows:

Part I   **General**
Regulation 1   Citation, commencement and revocation (see commentary to regulation 1 in Appendix 2)
Regulation 2   Interpretation (see commentary to regulation 2 in Appendix 2)

Part II   **Grant and withdrawal of approval**
Regulation 3   Approval of inspectors (see Section 5.5)
Regulation 4   Designation of bodies to approve inspectors (see Section 5.5)
Regulation 5   Manner of approval or designation (see Section 5.5)
Regulation 6   Termination of approval or designation (see Section 5.5.2)
Regulation 7   Lists of approvals and designations (see Section 5.5.1)

Part III   **Supervision of work by Approved Inspectors**
Regulation 8   Initial notice (see Section 5.6.5)
Regulation 9   Amendment notice (see Section 5.6.9)
Regulation 10   Independence of Approved Inspectors (see Section 5.6.4)
Regulation 11   Functions of Approved Inspectors (see Section 5.6.6)
Regulation 12   Energy rating (see Section 5.6.6)
Regulation 12A   Sound insulation testing (see Section 5.8.2)
Regulation 13   Approved Inspector's consultation with the Fire Authority (see Section 5.7.1)
Regulation 13A   Approved Inspector's consultation with the Sewerage undertaker (see Section 5.7.2)

Part IV   **Plans certificates**
Regulation 14   Form of plans certificate (see Section 5.6.7 and Figure 5.5)
Regulation 15   Grounds and period for rejecting plans certificate (see Section 5.6.8)
Regulation 16   Effect of plans certificate (see Section 5.6.6)

Part V   **Final certificates**
Regulation 17   Form, grounds and period for rejecting final certificate (see Section 5.8.4 and Figure 5.6)

Part VI   **Cessation of effect of initial notice**
Regulation 18   Events causing initial notice to cease to be in force (see Section 5.8.4)
Regulation 19   Cancellation of initial notice (see Section 5.6.5 and Figure 5.4, and Section 6.4 and Figure 6.1)
Regulation 20   Local authority powers in relation to partly completed work (see Sections 5.8.4 and 6.3)

Part VII	**Public bodies**
Regulation 21	Approval of public bodies (see Section 5.9)
Regulation 22	Public body's notice (see Section 5.9.1 and Figure 5.7)
Regulation 23	Public body's consultation with the fire authority (see Section 5.9)
Regulation 23A	Public body's consultation with the sewerage undertaker (see Section 5.9)
Regulation 24	Public body's plans certificate (see Section 5.9.2)
Regulation 25	Grounds and period for rejecting public body's plans certificate (see Section 5.9.2)
Regulation 26	Effect of public body's plans certificate (see Section 5.9.2)
Regulation 27	Public body's final certificate (see Section 5.9.3)
Regulation 28	Events causing public body's notice to cease to be in force (see Section 5.9.1)

Part VIII	**Certificates relating to deposited plans**
Regulation 29	Certificates given under section 16(9) of the Act (see Section 5.2.2 and commentary to regulation 29 in Appendix 2)

This regulation deals with the giving of certificates of conformity with building regulations by persons approved for that purpose. Such persons must be approved by a designated body in the same way as for an approved inspector. The regulations which apply to the approval and termination of approval of both Approved Inspectors and Designated Bodies also apply to the approval of approved persons and their designated bodies (see regulations 3 to 7 of the *Building (Approved Inspectors etc) Regulations 2000*)

Part IX	**Registers**
Regulation 30	Register of notices and certificates (see Section 5.4.11)

Part X	**Effect of contravening building regulations**
Regulation 31	Contravention of certain regulations not to be an offence

In most cases, contravening the *Building (Approved Inspectors etc) Regulations 2000* does not constitute an offence under section 35 of the *Building Act 1984*. The exceptions are in the cases of:

- Regulation 12 energy rating (see Section 5.6.6)
- Regulation 12A sound insulation testing (see Section 5.8.2)
- Regulation 20 local authority powers in relation to partly completed work (see Sections 5.8.4 and 6.3).

Part XI	**Miscellaneous provisions**
Regulation 32	Transitional provisions (see commentary to regulation 1(1) in Appendix 2)
Schedule 1	Revocation of Regulations (see Appendix 2)
Schedule 2	Forms
	Form 1 Initial notice (see Figure 5.3)

Form 2 Amendment notice (see Section 5.6.9)

Form 3 Plans certificate (see Figure 5.5)

Form 4 Combined initial notice and plans certificate (see Section 5.6.7)

Form 5 Final certificate (see Figure 5.6)

Form 6 Notice of cancellation by approved inspector (see Figure 6.1)

Form 7 Notice of cancellation by person carrying out work (see Figure 5.4)

Form 8 Notice of cancellation by local authority (see Section 5.6.5)

Form 9 Public body's notice (see Chapter 5 Figure 5.7)

Form 10 Public body's plans certificate (see Section 5.9.2)

Form 11 Combined public body's notice and plans certificate (see Section 5.9.1)

Form 12 Public body's final certificate (see Section 5.9.3)

Schedule 3   Grounds for rejecting an Initial Notice, an Amendment Notice, or a Plans Certificate Combined with an Initial Notice (see Section 5.6.5)

Schedule 4   Grounds for rejecting a Plans Certificate, or a Plans Certificate Combined with an Initial Notice (see Section 5.6.8)

Schedule 5   Grounds for rejecting a Final Certificate (see Section 5.8.4)

Schedule 6   Grounds for rejecting a Public Body's Notice, or a Combined Public Body's Notice and Plans Certificate (see Sections 5.9.1 and 5.9.2)

Schedule 7   Grounds for rejecting a Public Body's Plans Certificate or a Combined Public Body's Notice and Plans Certificate (see Section 5.9.2)

Schedule 8   Grounds for rejecting a Public Body's Final Certificate (see Section 5.9.3).

Application of the regulations

4.1 What type of work is covered by the regulations?

Put simply, the regulations apply to 'building work'. Since this term could have any number of meanings, it is closely defined in regulation 3 of the *Building Regulations 2000* (as amended) (referred to as the 2000 Regulations in the remainder of this chapter) and is explained in detail in Section 4.1.1. Building work (whether to a new or existing building) must comply with the applicable requirements of Schedule 1 to the 2000 Regulations (see regulation 4). These requirements are phrased in functional terms and cover such aspects of building design and construction as structural stability, safety in fire, resistance to dampness, ventilation, drainage, waste disposal, etc. They are given in full in Appendix 1. It is possible that satisfying one part of Schedule 1 can cause a contravention of another part (e.g. from a purely structural viewpoint, in some circumstances, external walls could be constructed of 190-millimetre-thick masonry; however, this form of construction would be unlikely to satisfy the weather-resistance requirements of Part C and the heat-loss requirements of Part L). Therefore, regulation 4(1)(b) prevents this by requiring that all parts of Schedule 1 be complied with (one requirement must not cause a contravention of another). When carrying out work to existing buildings it is not always necessary to bring the building up to the full requirements of all parts of Schedule 1; however, work to existing buildings must be carried out so that, when the work is complete, the building is not made worse in relation to compliance with the regulations than it was before the work commenced.

4.1.1 Meaning of building work

In the regulation 3(1) of the 2000 Regulations 'building work' means:

(1) the erection or extension of a building
(2) the provision or extension of a controlled service or fitting in or in connection with a building (but see regulation 1A below)

(3) the material alteration of a building, or a controlled service or fitting, as mentioned in paragraph 3(2) of the 2000 Regulations
(4) work required by regulation 6 (requirements relating to a material change of use)
(5) the insertion of insulating material into the cavity wall of a building
(6) work involving the underpinning of a building.

What is a 'building'?

In the regulations the term 'building' is defined as *'any permanent or temporary building but not any other kind of structure or erection'*. The reference to a building is taken to include part of a building.

This definition is narrower than that contained in section 121 of the *Building Act 1984*, the purpose being to restrict the scope of the Regulations to what are commonly thought of and referred to as buildings.

The Building Act definition is necessarily couched in very wide terms so as to provide the local authority with the necessary powers to deal with (e.g. dangerous structures and demolitions). In the past, doubt has been cast over the status of structures, such as residential park-homes and marquees. Provided that a residential park-home conforms to the definition given in the *Caravan Sites and Control of Development Act 1960* (as augmented by the *Caravan Sites Act 1968*) it is exempt from the definition of 'building' contained in the regulations, and according to the *Manual to the Building Regulations* a marquee is not regarded as a building.

4.1.2 Erection or extension of a building

Although the regulations do not attempt to define 'erection of a building', *section 123 of the Building Act 1984* gives a relevant statutory definition for 'erection of a building' which includes *'the reconstruction of a building,* [and] *the roofing over of an open space between walls or buildings'*. More specifically, certain building operations are *'deemed to be the erection of a building'* as follows:

'(a) *the re-erection of any building or part of a building when an outer wall has been pulled or burnt down to within ten feet of the surface of the ground adjoining the lowest storey of the building or of that part of the building,*
(b) *the re-erection of a frame building or part of a frame building when that building or part of a building has been so far pulled down or burnt down, as to leave only the framework of the lowest storey of the building or of that part of the building,*
(c) *the roofing over of an open space between walls or buildings.'*

In the cases mentioned, the work would be subject to control under the regulations where the conditions specified were met. Any work additional to the re-erection, etc. of the building (whether or not the building met the conditions above), such as the provision of new services or a change of use, would of course, be covered by the regulations.

4.1.3 Controlled services and fittings

The only services and fittings controlled by the regulations are those concerned with the provision, extension or material alteration of:

(a) bathroom fittings, hot water storage systems and sanitary conveniences under Part G
(b) drainage and waste disposal systems under Part H
(c) combustion appliances under Part J
(d) and the provision (including replacement) or extension of the following under Part L:
   (i) replacement external doors, windows, rooflights and roof windows
   (ii) space heating systems (and associated boilers, hot water pipes and hot air ducts, etc.)
   (iii) hot water systems (and associated boilers and hot water vessels, etc.)
   (iv) lighting systems
   (v) (*in non-domestic buildings*) hot water service pipes
   (vi) (*in non-domestic buildings*) air conditioning and mechanical ventilation systems (and associated chilled water and refrigerant vessels and pipes and air ducts).

Obviously, some of the items listed under (d) will already be controlled services or fittings because of the requirements of Parts G and J.

The extent of compliance is limited by the parts of the regulations referred to above and in the case of services and fittings in (d) there are further limitations applied by virtue of regulation 3(1)(A) where the services and fittings are located in dwellings (see below).

The extent of compliance where the work involves a material alteration to a controlled service or fitting is also governed by regulation 3(2) (see below). For example, if it was proposed to install sanitary fittings, such as a water closet, bath, shower, washbasin or kitchen sink in a building, the regulations would apply to the installation if it involved alterations to, or new connections to, a drainage stack or an underground drain. Simple replacement of a fitting not involving an extension to the drainage pipework would not be covered by the regulations.

4.1.4 The installation of replacement services and fittings in existing dwellings under Part L

Paragraph 3(1)(A) brings under the definition of building work the provision (including replacement) or extension of the following services and fittings in existing dwellings:

• a window, rooflight, roof window or door (with, including its frame, more than 50% of its internal face area glazed), or

- a space heating or hot water service boiler, or
- a hot water vessel.

However, the provision or extension, in existing dwellings, of other services or fittings in relation to which Part L imposes requirements, is not building work, unless the services and fittings in question are 'controlled' because they are also subject to Part G, H or J. For example, the installation of an energy-efficient lighting system in an existing dwelling is not building work and is not covered by the regulations.

Examples of the way in which the regulation requirements should be applied are given in Approved Document L1 and follow below. In general, Part L1 applies to replacement work on controlled services or fittings when:

- replacing old equipment with new identical equipment
- replacing old equipment with new but different equipment
- the work is solely in connection with controlled services or includes work on them.

Example 1: Replacement windows, doors, rooflights and roof windows
The requirements only apply to the provision of a complete new window or door installation (i.e. the fixed frame as well as the moving element) therefore repair work to parts of the element, such as:

- replacing broken glass
- replacing a sealed double glazing unit or
- replacing rotten framing members

is not covered by the requirement.

As well as satisfying the insulation recommendations of Part L, Approved Document L also states that the glazing should comply with Part N *Glazing – Safety in relation to impact, opening and cleaning* and should not be made worse in relation to Part B *Fire* (in terms of means of escape), Part F *Ventilation* (in terms of the provision of fresh air for the occupants) and Part J *Combustion appliances and fuel storage systems* (in terms of the provision of combustion air for the appliance).

Example 2: Replacement heating boilers and hot water vessels
Replacement boilers must be capable of achieving a satisfactory SEDBUK (Seasonal Efficiency of Domestic Boilers in the UK) efficiency and must have suitable controls to allow a satisfactory SEDBUK efficiency to be achieved in practice. Replacement boilers must be tested and commissioned, and instructions must be provided to facilitate energy-efficient operation.

When replacing hot water vessels, new equipment should be provided which would satisfy the requirements for a new dwelling.

4.1.5 Material alteration of a building

Building work which constitutes of a 'material alteration' to a building is subject to control under the regulations. The definition of material alteration in regulation 3(2) is phrased in general terms and can cause problems of interpretation. Each case needs to be judged on its own merits against the general principles stated in regulation 3(2). Thus, an alteration is deemed to be a '*material alteration*' and so subject to control under the regulations if the alteration would cause the existing building, at any stage, to:

(1) not meet certain specified requirements (the '*relevant requirements*') of the regulations which it did meet before the alterations were commenced, or
(2) be made worse in relation to compliance with the relevant requirements, if the building was already not in compliance before the alterations were commenced.

The '*relevant requirements*' are:

- Part A (Structure)
- In Part B (Fire safety), paragraph B1 (means of warning and escape), paragraph B3 (internal fire spread – structure), paragraph B4 (external fire spread) and paragraph B5 (access and facilities for the fire service)
- Part M (access and facilities for disabled people).

In general, it is not necessary to bring the existing building up to current regulation standards, however, it should not be made worse as a result of the works of alteration when measured against the relevant requirements. Special note should be made of the phrase '*at any stage*' in the first line of regulation 3(2) (see Appendix 1) since this indicates that the regulations apply not only to completed work but also to partially completed work and, indeed, to work in progress. This has obvious implications for structural alterations (unsatisfactory temporary support) and alterations that affect a means of escape (where it might be blocked by the work of alteration).

It is often difficult to decide whether or not the work being carried out constitutes a material alteration (and is therefore subject to control) or whether it is merely repairs (and therefore not subject to control). No definition of 'repair' appears in the regulations, however, the *Manual to the Building Regulations* makes it clear that works of repair (replacement, redecoration, routine maintenance and making good) do not come under the definition of material alteration. Additionally, the 2002 edition of the Office of the Deputy Prime Minister's (ODPM's) *Building Regulations Explanatory Booklet* indicates that repairs of a minor nature, such as:

- replacing roofing tiles with the same type and weight of tile
- replacing the felt to a flat roof
- repointing brickwork or
- replacing floorboards

are not deemed to be material alterations.

More significant 'repair' work, such as:

- removing a substantial part of a wall and rebuilding it (not including a garden or boundary wall)
- underpinning a building

- installing a new flue or flue liner
- forming a new opening in a structural wall

are material alterations and therefore subject to control.

In the case of re-roofing works the situation may not be clear-cut. For example, if the new tiling or roofing material is substantially heavier than the existing material the regulations may apply – it will depend on the extent of the change in weight. In the 2004 edition of Approved Document A (Structure), it is suggested that a significant change in roof loading is when the self-weight loading on the roof is increased by more than 15%. Unfortunately, a significant decrease in roof dead loading (which might also cause the regulations to be activated) is not so clearly defined. However, if the roof is thatched, or is to be thatched where previously it was not, then the regulations will apply.

4.1.6 Material change of use: interpretation

Not all use changes are controlled by the regulations. The main reasons for requiring control of some building use changes and not others are usually related to a perceived increase in the risk to the health and safety of the occupants, due either to greater fire hazard being created and/or increased loading being placed on the building fabric, as a result of the change.

Use changes covered by the regulations are defined in regulation 5 as '*material changes of use*' and are those where:

(a) the building is used as a dwelling, hotel, boarding house, shop, institution or public building where previously it was not
(b) the building contains a flat where previously it did not
(c) the building is no longer exempt under Schedule 2, Classes I to VI, where previously it was
(d) a building containing at least one dwelling is converted to provide a greater or lesser number of dwellings than it did previously
(e) the building contains a room for residential purposes, where previously it did not, or where it contains at least one such room, it is converted to contain a greater or lesser number of such rooms than before.

In the above list:

- The term dwelling includes a flat.
- With regard to hotels and boarding houses, the *Fire Precautions Act 1971* requires many of these premises to carry a fire certificate. Decided case law would indicate that the use of premises for bed-sitting accommodation is materially different from hotel use (by virtue of the fact that hotel populations are normally transient whereas bed-sitting rooms are usually people's homes).
- Regulation 2 defines dwelling, flat, institution, shop, public building and room for residential purposes (see Appendix 1).

- Schedule 2 (see Appendix 1) contains a list of small buildings and extensions that are exempt from control under the regulations. The effect of regulation 5 is to make it a material change of use if a building was originally constructed so as to be exempt under Schedule 2 (Classes I to VI) and it is subsequently used for another purpose. Interestingly, Class VII (Extensions) is not included in this regulation. Therefore, for example, it would appear that the conversion of a small detached domestic garage to an office would constitute a material change of use, whereas the change of use of an attached conservatory to an office at the same property would not (although it might require planning approval).

4.1.7 Material change of use: application of the regulation requirements

Regulation 6 lists those parts of Schedule 1 which are applicable to the various types of material change of use described in regulation 5 and listed above. Table 4.1 sets out the requirements that apply to each use change.

The items listed in 6(1)(a) apply to all changes of use where the change affects the whole of a building. The items listed in 6(1)(b) to 6(1)(g) apply additional parts of Schedule 1 to specified use changes (e.g. Part E, *Resistance to the passage of sound*, only applies to changes which result in the creation of dwellings, flats, hotels, and boarding houses and rooms for residential purposes).

Where a change of use affects only part of a building, in most cases it is only that part which must comply with the listed parts of Schedule 1 (i.e. not the whole building). The exceptions are in relation to:

- buildings exceeding 15 metres in height (see 6(1)(c), where the whole building must comply with B4(1) (*External fire spread – walls*))
- a change of use to a hotel, boarding house, shop, institution or public building where the whole building must comply with M1(a) (*Access and use*) regarding access to the part being converted.

4.1.8 The insertion of insulating material into the cavity wall of a building

The insertion of cavity wall insulation is building work and is controlled by the regulations. It must comply with the requirements of paragraph C4 (*Resistance to weather and ground moisture*) and paragraph D1 (*Cavity insulation*). Paragraph 13(3) of the 2000 Regulations (see Appendix 1) contains details of additional information which must accompany a building notice where there is an intention to insert insulating material into the cavity walls of a building. This is to ensure that the relevant work is of a suitable specification and is installed by a suitably qualified and experienced person. Since regulation 3(1)(e) refers only to the insertion

Table 4.1 Material change of use – application of the regulation requirements

Section of regulation 6	Change of use	Applicable requirements
(1)(a)[1]	• Building use changed to dwelling, hotel, boarding house, shop, institution or public building • Building use changed so as to contain a flat • Building no longer exempt under Schedule 2, Classes I to VI, where previously it was • Building containing at least one dwelling is converted to provide a greater or lesser number of dwellings than before • Building contains a room for residential purposes, where previously it did not, or where it contains at least one such room, is converted to contain a greater or lesser number of such rooms than before	B1 (means of warning and escape) B2 (internal fire spread – linings) B3 (internal fire spread – structure) B4(2) (external fire spread – roofs) B5 (access and facilities for the fire service) C2(c) (interstitial and surface condensation) F1 (ventilation) G1 (sanitary conveniences and washing facilities) G2 (bathrooms) H1 (foul water drainage) H6 (solid waste storage) J1 to J3 (combustion appliances) L1 (conservation of fuel and power – dwellings) L2 (conservation of fuel and power – buildings other than dwellings) P1 and P2 (electrical safety)
(1)(b)[1]	• Building use changed to hotel, boarding house, institution or public building; or • Building no longer exempt under Schedule 2, Classes I to VI, where previously it was	A1 to A3 (structure)
(1)(c)[1]	• Building of any use change exceeding 15 metres in height	B4(1) (external fire spread – walls)
(1)(cc)	• Building use changed to dwelling, hotel, boarding house or institution • Building use changed so as to contain a flat • Building containing at least one dwelling is converted to provide a greater or lesser number of dwellings than before • Building contains a room for residential purposes, where previously it did not, or where it contains at least one such room, is converted to contain a greater or lesser number of such rooms than before	C1(2) (resistance to contaminants)

- Building no longer exempt under Schedule 2, Classes I to VI, where previously it was, where the material alteration provides new residential accommodation

(1)(d)[1]
- Building use changed to dwelling

(1)(e)[1]
- Building use changed to dwelling, hotel or boarding house
- Building use changed so as to contain a flat
- Building containing at least one dwelling is converted to provide a greater or lesser number of dwellings than before
- Building contains a room for residential purposes, where previously it did not, or where it contains at least one such room, it is converted to contain a greater or lesser number of such rooms than before

(1)(f)[1]
- Building use changed to public building consisting of or containing a school

(1)(g)[1]
- Building use changed to hotel, boarding house, shop, institution or public building

(2)(a) and (b)[2]
Change of use affecting only part of a building in cases (1)(a), (b), (d), (e) or (f) above

(2)(c)[2]
Change of use affecting only part of a building in case (1)(c)

(2)(d)[2]
Change of use affecting only part of a building in case (1)(g)

C2 (resistance to moisture)

E1 to E3 (resistance to the passage of sound)

E4 (acoustic conditions in schools)

M1 (access and use)

Requirements referred to in cases (1)(a), (b), (d), (e) or (f) above applied only to the part of the building being changed

B4(1) (external fire spread – walls) applied to the whole building (not just the part being changed)

(i) M1 (access and use) applied to the being changed and to any sanitary conveniences provided in or in connection with the part being changed

(ii) Whole building complies with requirement M1(a) of Schedule 1 to the extent that reasonable provision is made to provide either suitable independent access to the part being changed or suitable access through the building to that part

Note 1: Sections (1)(a) to (1)(g) apply to a change of use that affects the whole building.
Note 2: Sections (2)(a) to (2)(d) apply to a change of use that affects only part of a building.

of insulation into the cavity in a wall it would appear that the provision of external wall insulation is not building work and is therefore not subject to control. It is of course possible that such work might be construed as a material alteration to the building since it might possibly affect the structure of the building under Part A or the external fire spread under paragraph B4.

4.1.9 Underpinning of a building

Work involving the underpinning of a building is controlled by the regulations. The regulations are applied so as to ensure that any movement of the building is stabilised (see Part A *Structure*, paragraph A1 *Loading*). Regard will also need to be given to the effect on any sewers or drains near the work, therefore reference should also be made to Part H, *Drainage and waste disposal*, paragraph H1 *Foul water drainage*, paragraph H3 *Rain water drainage* and paragraph H4 *Building over sewers*.

4.2 Are there any exemptions?

Although the Building Act 1984 and the 2000 Regulations apply to building work in England and Wales, taken together these two pieces of legislation exempt certain classes and uses of buildings, and many categories of work from control. These include:

- Crown buildings.
- Educational buildings.
- Buildings belonging to statutory undertakers and other public bodies.
- Certain small buildings and extensions.

4.2.1 Crown buildings

The Building Regulations do not apply to premises which are occupied by the Crown. It is an established rule of statutory interpretation that the Crown is not bound by an Act of Parliament except by express provision or necessary implication. Therefore, an Act of Parliament must specifically state that the Crown is covered by the provisions in order that it be bound by them and in fact, there is a provision in section 44 of the *Building Act 1984* to apply the substantive requirements of the regulations to Crown buildings but this has never been activated.

According to the *Building Act 1984*, a Crown building is defined as '*a building in which there is a Crown interest or a Duchy interest*'. This definition necessitates the following additional definitions.

Crown interest means '*an interest belonging to Her Majesty in right of the Crown, or belonging to a government department, or held in trust for Her Majesty for the purposes of a government department*'.

Duchy interest means '*an interest belonging to Her Majesty in right of the Duchy of Lancaster, or belonging to the Duchy of Cornwall*'.

Examples of Crown buildings include not only the Royal Palaces, the Houses of Parliament, 10 Downing Street, etc., but also all Government offices (such as local Job Centres) across England and Wales.

Over the years a number of bodies have lost Crown immunity. These include:

- *Health Service Premises*: Under the provisions of section 60 of the *National Health Service and Community Care Act 1990* health service bodies are no longer regarded as the servant or agent of the Crown in respect of land over which they have powers of disposal or management, or which is otherwise used or occupied by them. Subsection (7) of section 60 defines *Health Service Bodies* in relation to England and Wales as a Family Health Services Authority, the Dental Practice Board and the Public Health Laboratory Service Board. In practice this covers regional, district and special health authorities, and means that health service buildings are now subject to the full substantive and procedural provisions of building, planning and fire precautions legislation enforceable by local authorities.
- *The Metropolitan Police*: Although no longer regarded as servants or agents of the Crown, the Metropolitan Police Authority has been exempted from having to comply with the procedural requirements of the building regulations, using the powers available under section 5 of the Building Act 1984 (*exemption of public bodies from the procedural requirements and enforcement of building regulations*). However, it is still required to comply with the substantive or technical requirements of the Regulations. As an exempt body the Metropolitan Police Authority is also exempt from enforcement procedures by local authorities. Instead, the Metropolitan Police Authority as a '*Public Body*' is bound by the provisions of the *Building Act 1984, section 54 (Supervision of their own work by public bodies)* and by *Part VII (Public Bodies)* of the *Building (Approved Inspectors etc) Regulations 2000 (SI 2000/2532)* (as amended). See Chapter 5 for details of the procedures regarding public bodies.

4.2.2 Educational buildings

As a result of the repeal of regulation 8 of the *Education (Schools and Further and Higher Education) Regulations 1989*, maintained schools in England ceased to have exemption from the Building Regulations from 1 April 2000. A similar situation has existed in Wales since 1 January 2002 following the passing of the *Education (Schools and Higher and Further Education) (Amendment) (Wales) Regulations 2001*. As a result, building works at schools are now treated in the same way as in other user groups and are subject to normal building control procedures. This is a change in the approval process meaning that building regulation submissions in respect of work to maintained schools now has to be made to the appropriate building control body (Local Authority or Approved Inspector), but does not affect the standards applicable to schools. The change does not affect in any way the status of the *Education (School Premises) Regulations 1999*, which continues to apply to all schools. These regulations cover general standards of provision of facilities, such as:

- In day schools, accommodation for washrooms, medical purposes, staff, cloakrooms, canteens, etc.
- In boarding schools, accommodation for sleeping, washing (including bathrooms), living (for study outside school hours and for social purposes), preparing and consuming meals, medical purposes, staff and storage.

They also cover general constructional requirements, such as:

- Structural stability.
- Weather protection.
- Means of escape in case of fire and other health, safety and welfare issues.
- Acoustics, lighting, heating and ventilation.
- Water supplies and drainage.

For many years, constructional standards for schools have been set by the Department for Education and Science (DfES) in England and the National Assembly for Wales in Wales, the most recent ones being the 1997 Constructional Standards. Virtually all of the requirements of these standards for school buildings have now been incorporated into the building regulation Approved Documents, which sometimes refer to DfES *Building Bulletins* as alternatives to the normal Approved Document guidance. For example:

- For ventilation, Approved Document F refers to Building Bulletin 87 as an appropriate design standard for schools.
- For acoustics, Approved Document E (*Resistance to the passage of sound*) refers the reader to Building Bulletin 93 (*The acoustic design of schools*) for guidance on acoustic conditions and disturbance by noise.

Care should be taken to use the most recent edition of these alternative sources of guidance since they are regularly updated by DfES.

Further information on alternative sources of guidance can be obtained from the School Premises Team, Department for Education and Skills, Caxton House, Room 762, 6–12 Tothill Street, London SW1H 9NA; Tel.: 020 7273 6023; E-mail: premises.schools@dfes.gsi.gov.uk.

4.2.3 Buildings belonging to statutory undertakers and other public bodies

Under section 4 of the Building Act 1984, a building belonging to a statutory undertaker, the UK Atomic Energy Authority, or the Civil Aviation Authority is exempt from the application of building regulations, provided that the building in question is held or used by them for the purposes of their undertaking.

'Statutory undertaker' as defined in section 126 of the *Building Act 1984* (as amended) means '*persons authorised by an enactment or statutory order to construct, work or carry on a railway, canal, inland navigation, dock harbour, tramway or other public undertaking but does not include a universal service provider (within the meaning of the Postal Services Act 2000), the Post Office company (within the meaning of Part IV of that Act) or any subsidiary or wholly-owned subsidiary (within the meanings given by section 736 of the Companies Act 1985) of the Post Office company*'.

From this definition it is clear that Post Offices (i.e. the actual high-street 'shops' where the public resort for postal services) are no longer regarded as statutory undertakers within the meaning of section 126, consequently they are now required to comply with the building regulations including submissions of work to the appropriate building control body. Interestingly, Royal Mail (which deals with the collection, sorting and delivery of mail) is still regarded as a Crown body and is therefore exempt from compliance with the building regulations.

The status of statutory undertakers has been complicated by the fact that a number of former public bodies are now in private hands. This has meant that it has been necessary to pass additional legislation in order to clarify the status of some of these bodies. As a consequence, the following bodies are deemed to be statutory undertakers:

- Public gas suppliers (see sections 67(1), 67(3) and 67(4) of the *Gas Act 1986*).
- Electricity suppliers (see section 112(4) of the *Electricity Act 1989*).
- The National Rivers Authority (see section 190(1) of the *Water Act 1989*).
- Water and sewerage undertakers (see section 190(1) of the *Water Act 1989*).

The building of the statutory undertaker must be one which is held or used by them for the purposes of the undertaking, therefore the exemption from the application of building regulations granted by virtue of section 4 of the *Building Act 1984* is subject to the following exceptions, in respect of which building regulations do apply:

(i) a house
(ii) a building used as offices or showrooms unless:
   (a) it forms part of a railway station, or
   (b) in the case of the Civil Aviation Authority, it is on an aerodrome owned by the Authority.

A further class which enjoys an exemption under section 4 is 'relevant airport operators' as defined in section 57 of the *Airports Act 1986*. Section 4 applies in relation to a relevant airport operator as it applies to a statutory undertaker, subject to the following variations:

(i) hotels are not exempt from compliance with the regulations
(ii) offices and showrooms are exempt from compliance with the regulations, if they are on any airport to which Part V of the Airport Act 1986 applies.

4.2.4 Certain small buildings and extensions

The regulations do not apply to the erection of any building set out in Classes I to VI, or to extension work in Class VII of Schedule 2 to the Building Regulations 2000. Full details of the Classes covered are given in Appendix 1. They include:

Class I   **Buildings controlled under other legislation** – such as:

- A building, the construction of which is controlled under the Explosives Acts 1875 and 1923 (e.g. an explosives store at a quarry).
- Any building erected on a site licensed under the Nuclear Installations Act 1965 (other than a building containing a dwelling or a building used for office or canteen accommodation).
- Any building scheduled under section 1 of the Ancient Monuments and Archaeological Areas Act 1979.

The point of the Class I exemptions is that where such specialist legislation exists to control certain risks there is no point in having this overlaid with more general (and possibly conflicting) legislation.

Class II   **Buildings not frequented by people** – such as detached buildings into which people do not normally go, or only use intermittently for inspecting or maintaining fixed plant or machinery (e.g. a detached boiler or plant room).

Such buildings are only exempt where they are at least one-and-a-half times their own height from the site boundary or from any other buildings which are frequented by people.

The Class II exemptions apply because of the relatively low risk to health and safety associated with such buildings.

Class III   **Greenhouses and agricultural buildings** – these must be used for agriculture or the keeping of animals ('agriculture' includes horticulture, fruit growing, the growing of plants for seed and fish farming).

Additionally, for the exemption to apply:

- no part of the building must used as a dwelling
- the building must be at least one-and-a-half times its height from any building containing sleeping accommodation
- the building must be provided with a fire exit which is not more than 30 metres from any point in the building.

The exemption does not apply to a greenhouse or a building used for agriculture if the principal purpose for which it is used is retailing, packing or exhibiting (such as at a garden centre).

Class III buildings are considered to be low risk provided that the conditions are met, since they are only occupied by small numbers of people under normal daytime working conditions.

Class IV **Temporary buildings** – buildings intended to remain where they are erected for 28 days or less, such as exhibition stands.

Class V **Ancillary buildings** – site buildings, often with limited life. Specifically, a building:

- on a site intended to be used only in connection with the disposal of buildings or building plots on that site
- on the site of construction or civil engineering works, intended to be used only during the course of the works and containing no sleeping accommodation
- erected for use on the site of and in connection with a mine or quarry. The building must not contain a dwelling or be used as an office or showroom.

Class VI **Small detached buildings** – three categories of buildings are exempted because of their small size and specialist and/or occasional use. These are:

- A detached single storey building of up to 30 square metres floor area not containing sleeping accommodation. For the exemption to apply, the building must either be situated more than 1 metre from the boundary of the curtilage or be constructed substantially from non-combustible material.
- A detached building of up to 30 square metres floor area designed and intended to shelter people from the effects of nuclear, chemical or conventional weapons, and not used for any other purpose. Since such buildings are invariably constructed below ground there is a risk of undermining the foundations of other buildings on the site. Therefore, for the exemption to apply the excavation for the shelter must be no closer to any exposed part of another building or structure than a distance equal to the depth of the excavation plus 1 metre.
- A detached building with a floor area not exceeding 15 square metres, containing no sleeping accommodation, such as a garden shed.

Class VII **Extensions** – a small (floor area not exceeding 30 square metres) ground level extension to a building consisting of a conservatory, porch, covered yard or covered way, or a carport open on at least two sides. However, in the case of a conservatory or porch which is wholly or partly glazed, the glazing must satisfy the requirements of Part N of Schedule 1.

The building regulations have no application at all to any work done to or in connection with buildings in Classes I to VII provided that, in the case of Classes I to VI, the work does not involve a material change of use. Regulation 5(f) makes it a material change of use if a building was originally constructed so as to be exempt under Schedule 2 (Classes I to VI) and it is subsequently used for another purpose. It should be noted that in the case of Class VII conservatories or porches, in order for the exemption to apply the glazing must satisfy the requirements of Part N (*Glazing – Safety in relation to impact, opening and cleaning*). This raises and interesting conundrum since exempt work is not notifiable to the Local Authority or Approved Inspector, therefore no check would be carried out to confirm that the glazing did comply. However, it does make enforcement action possible where a breach of building regulations can be shown. Proposals to bring conservatories back under control are currently out for consultation.

4.2.5 Repairs to buildings

Whilst it is certain that simple repairs to buildings are not covered by the regulations, the extent of the exemption can only be decided by default (i.e. does the work come under the definition of material alteration or not) since no specific definition of repair appears in the regulations. No definition of 'repair' appears in the regulations, however, the *Manual to the Building Regulations* makes it clear that works of repair (replacement, redecoration, routine maintenance and making good) do not come under the definition of material alteration. Additionally, the 2002 edition of the ODPM's *Building Regulations Explanatory Booklet* indicates that repairs of a minor nature are not deemed to be material alterations. For more details on repairs and alterations see Section 4.1.5.

4.3 Links with other legislation

In this discussion of the application of the building regulations it is important to realise that there are a number of key Acts of Parliament and Statutory Instruments that are closely linked to the regulations, and can or must be satisfied when compliance with the regulations is required to be sought. The purpose of this section is to make the reader aware of these statutory requirements, since some of them may be effective even where the building regulations do not apply to a proposal. In practice, there is a great deal of legislation which affects building development so, of necessity, the information given is not a comprehensive list of the Acts of Parliament and regulations which apply in all circumstances, but a summary of some of those most commonly encountered.

4.3.1 The Building Act 1984

This Act is pivotal to the operation of the building control system in England and Wales, and is described in detail in Section 3.1, and is touched on in most chapters of this book. It contains the powers to make building regulations and the rules which affect administration and enforcement of the regulations. Therefore, it runs through and behind every aspect of the regulations and without it our control system would not exist.

The linked powers

An example of a direct link which is triggered by the making of a regulation submission to a Local Authority or the giving of an initial notice by an Approved Inspector concerns the so-called 'linked powers'.

Under section 16 the local authority must pass the deposited plans unless they are defective or show a contravention of the regulations. In the past, deposit of plans for the purposes of building regulation compliance triggered off a number of 'linked powers', other sections of the Building Act which came into effect on deposit of plans under the regulations, and the plans could be rejected if they did not show compliance with these other provisions.

In the 20 years since the Act came into force most of these linked powers have been replaced by changes in the regulations themselves. The only remaining section relates to the provision for the occupants of a house of a 'supply of wholesome water sufficient for their domestic purposes'. This may be found in section 25 of the *Building Act 1984*. Full details of the application of this section may be found in Section 5.2.2.

Additionally, there are other parts of the Building Act which may affect building proposals even if the building regulations do not apply. These are referred to in Part III of the Act (sections 59 to 83). The most commonly encountered sections cover the powers of the local authority to control:

- dangerous and defective premises, sections 76 to 79
- the demolition of buildings, sections 80 to 83
- means of escape in case of fire from certain high buildings, section 72
- the raising of chimneys if overreached by building work on an adjoining building, section 73.

Dangerous and defective premises

Section 77 of the Building Act empowers a local authority to deal with a building or structure which is in a dangerous condition or is overloaded. The normal procedure is for a local authority to apply to a magistrates' court for an order requiring the owner to carry out remedial works, or at his option to demolish the building or structure and remove the resultant rubbish. The court may also restrict the use of all or part of the building if the danger arises from overloading. In default, the local authority may carry out the necessary works to render the building safe and recover the expenses incurred from the owner who may also be liable to a fine on summary conviction.

Where there is insufficient time to go through the 'normal' procedure outlined above, section 78 empowers the local authority to take immediate action in an emergency to remove the danger. The owner should be notified where practicable. The local authority may recover reasonable expenses incurred in taken action unless the magistrates' court considers that they might reasonably have proceeded under section 77. In this case the owner of the building may be entitled to recover compensation from the local authority.

Where a building is in such a ruinous or dilapidated condition or is so neglected as to be seriously detrimental to the amenities of the neighbourhood, section 79 allows the local authority to serve notice on the owner requiring him to repair or restore the building, or at his option to demolish it and clear the site.

The demolition of buildings

The act of demolition of a building is not covered by the regulations (although it may constitute development under the planning system). Nevertheless, demolition processes can result in the exposure of drains and other services, and can result in dangerous conditions being created on sites. Therefore, local authorities are given powers under sections 80 to 83 of the Building Act to control demolitions. Under section 80 a person who intends to demolish the whole or part of a building must notify the local authority, the occupier of any adjacent building and the gas or electricity authorities of the intention to demolish. The person must also comply with any requirements which the local authority may impose under section 82. These procedures do not apply to the demolition of:

- an internal part of an occupied building where it is intended that the building should continue to be occupied
- a building with a volume (ascertained by external measurement) which does not exceed 1750 cubic feet (50 cubic metres)
- a greenhouse, conservatory, shed or prefabricated garage which forms part of a larger building
- an agricultural building unless it is contiguous to a non-agricultural building or a building which falls within the preceding paragraph.

The local authority may serve a notice on the person undertaking the demolition to carry out any or all of the following works:

- shore up any adjacent buildings
- make weatherproof any surfaces of an adjacent building exposed by the demolition
- repair and make good any damage caused to any adjacent building as a result of the works of demolition
- remove material and rubbish resulting from the demolition and clear the site
- disconnect and seal, and/or remove any sewers or drains in or under the building
- make good the surface of the ground disturbed by removal, etc. of drains and/or sewers
- make arrangements with the gas, electricity and water authorities for the disconnection of supplies
- make suitable arrangements with the Fire Authority (and the Health and Safety Executive, if necessary) with regard to the burning of structures or materials on site
- take such steps in connection with the demolition as are necessary for the protection of the public and the preservation of public amenity.

Means of escape in case of fire from certain high buildings
Section 72 enables a local authority, after consultation with the fire authority, to require adequate means of escape in an existing *or* proposed building which has more than two storeys and contains floors exceeding 20 feet above the surface of the street or ground on any side of the building. Not all building types are covered by this section since it refers only to those which are:

- let into flats or tenement dwellings
- used as inns, hotels, boarding houses, hospitals, nursing homes, boarding schools, children's homes or similar institutions
- used as restaurants, shops, stores or warehouses, where there are upper floors which contain sleeping accommodation for people employed on the premises.

The local authority is empowered to serve a notice on the owner or developer (in the case of a proposed building), to carry out those works which are necessary in order to provide an adequate means of escape. Regarding proposed buildings, section 72 does not apply if paragraph B1 of the Building Regulations imposes a requirement. Additionally, for existing buildings, the local authority cannot apply the provisions of section 72 to a building which has a valid fire certificate provided under the *Fire Precautions Act 1971* or is a workplace subject to the provisions of the *Fire Precautions (Workplace) Regulations 1997*. Over the years the effectiveness of section 72 has been reduced by the passing of other more comprehensive legislation and it is only applicable now to existing flats or tenements. Even here, the Housing Act 1985 already covers means of escape in existing houses in multiple occupation.

The raising of chimneys if overreached by building work on an adjoining building
Section 73 of the Building Act is operative where a person decides to erect or raise the height of a building above the level of the chimneys or flues within 6 feet of or in the party wall of a neighbouring building. The local authority may serve a notice on that person requiring him to raise the chimneys of the neighbouring building. The notice may also require the owner or occupier of the neighbouring building to allow the work to be carried out or to carry out the work himself if he so elects. It should be noted that this section does not apply to Inner London.

4.3.2 Local Acts of Parliament

Although the Building Act 1984 attempted to rationalise the main controls over buildings, there are in fact a great many pieces of local legislation with the result that many local authorities have special powers relevant to building control.

Where a local Act is in force, its provisions must also be complied with, since many of these pieces of legislation were enacted to meet local needs and perceived deficiencies in national legislation. The Building Regulations make it clear that local enactments must be taken into account.

With the growth and development of building regulation control over fire precautions in particular, it is likely that most of the current local legislation is now outdated or has been superseded by the Building Regulations. In fact, some local enactments already contain a statutory bar which gives precedence to building regulations.

Local authorities are obliged by section 90 of the Building Act to keep a copy of any local Act provisions and these must be available for public inspection free of charge at all reasonable times.

A full list of Local Acts of Parliament may be found in Appendix 3 to this book where it will be seen that the most common local provisions relating to building control are:

- *Special fire precautions for basement garages or for large garages*: The usual provision is that if a basement garage for more than three vehicles or a garage for more than 20 vehicles is to be erected, the local authority can impose access, ventilation and safety requirements.
- *Fire precautions in high buildings or for large storage buildings*: There must be adequate access for the fire brigade in certain high buildings. A high building is one in excess of 18.3 metres and the local authority must be satisfied with the fire precautions and may impose conditions (e.g. fire alarm systems, fire brigade access, etc.). Large storage buildings in excess of 14,000 cubic metres are required to be fitted with sprinkler systems by some local Acts. (In many cases these requirements have been superseded by similar provisions in Approved Document B, Fire Safety, which supports Part B of Schedule 1 to the Building Regulations 2000).
- *Extension of means of escape provisions*: Section 72 of the Building Act 1984 (described above) is a provision under which the local authority can insist on the provision of means of escape where there is a storey which is more than 20 feet above ground level in certain types of buildings (e.g. hotels, boarding houses, hospitals, etc.). Local enactments replace the 20 feet by 4.5 metres and make certain other amendments to the national provisions.
- *Drainage systems*: In some cases, local legislation requires that every building must have separate foul and surface water drainage systems. Requirement H5 (*Separate systems of drainage*) in Part H of Schedule 1 to the Building Regulations 2000, came into force on 1 April 2002. In view of this the *Building (Repeal of Provisions of Local Acts) Regulations 2003* came into force on 1 March 2004. The effect of these Regulations is to repeal certain sections of the Acts indicated in Table 4.2, since they duplicate the provisions of H5 regarding the provision of separate systems of drainage.
- *Safety of stands at sports grounds*: In many areas, local Acts impose controls over the safety of stands at sports grounds. Again much of this local legislation has been largely superseded by the provisions of the Fire Safety and Safety of Places of Sport Act 1987.

Table 4.2 Repealed sections of Local Acts of Parliament

Local Act	Reference	Extent of repeal
East Ham Corporation Act 1957	1957 c xxxvii	Section 38(2)
West Yorkshire Act 1980	1980 c iv	Section 50
South Yorkshire Act 1980	1980 c xxxvii	Section 39
Staffordshire Act 1983	1983 c xviii	Section 18
County of Lancashire Act 1984	1984 c xxi	Section 20
Leicestershire Act 1985	1985 c xvii	Section 30

4.3.3 The Workplace (Health, Safety and Welfare) Regulations 1992

Made under the *Health and Safety at Work etc. Act 1974* the *Workplace (Health, Safety and Welfare) Regulations 1992* (Workplace Regulations) implement provisions of European Workplace Directive 89/654/EEC. A general duty is placed on employers to ensure that workplaces comply with the requirements of the regulations including provisions for:

- environmental measures, such as ventilation, temperature control, lighting levels and adequacy of room dimensions
- general welfare including cleanliness, sanitary conveniences and washing facilities
- safety measures, such as the use of windows, doors, stairs, ladders and ramps, etc.

The Health and Safety Executive is responsible for enforcing the Workplace Regulations. The mechanism for this is by means of the service of an improvement notice on the owner or occupier, requiring remedial measures in the building. Under section 23(3) of the *Health and Safety at Work 1974* such an improvement notice may not be served if the required remedial measures are more onerous than those necessary to secure compliance with current building regulations. Table 4.3 lists those parts of the building regulations which will satisfy the corresponding Workplace Regulations in order to prevent the service of an improvement notice. Therefore, there is a connection with the Building Regulations in that new buildings which follow the guidance in the Approved Documents will, in most cases, satisfy the specified requirements of the Workplace Regulations which, of course, apply to workplaces when they are being used. In this context the Workplace Regulations may also be applied to the common parts of flats and similar buildings if people, such as cleaners and caretakers, are employed to work in these common parts. Where the requirements of the Building Regulations listed in Table 4.3 do not apply to dwellings, the provisions may still be required as indicated above in order to satisfy the Workplace Regulations.

Table 4.3 Satisfying the Workplace (Health, Safety and Welfare) Regulations 1992

Column 1	Column 2	Column 3
Building Regulation requirement	**Relevant Workplace Regulation**	**Details of the Workplace Regulation referred to in Column 2**
Requirement F1: Means of ventilation	Regulation 6(1)	Requirements for ventilation
Requirement K1: Stairs, ladders and ramps	Regulation 17	Permanent stairs, ladders and ramps on pedestrian traffic routes within the workplace premises, including those used to give access for maintenance to parts of the workplace premises
Requirement K2: Stairs, ladders and ramps	Regulation 13	Guarding with regard to the requirements for protection from the risk of falling a distance likely to cause personal injury
Requirement K3(2): Stairs, ladders and ramps	Regulation 17	Design of loading bays
Requirement K4: Protection from collision with open windows, etc.	Regulation 15(2)	Requirements for projecting windows, skylights and ventilators
Requirement K5: Protection against impact from and trapping by doors	Regulation 18	Requirements for doors and gates
Requirement N3: Safe opening and closing of windows, etc.	Regulation 15(1)	Requirements for opening, closing or adjusting windows, skylights and ventilators
Requirement N4: Safe access for cleaning windows, etc.	Regulation 16	Requirements for cleaning windows and skylights, etc.

4.3.4 The Fire Precautions (Workplace) Regulations 1997

The *Fire Precautions (Workplace) Regulations 1997* (as amended by the *Fire Precautions (Workplace) (Amendment) Regulations 1999*) require the provision of minimum fire safety standards in workplaces and impose duties on employers and on others in control of places of work with regard to the provision of minimum fire safety standards. They apply to any place of work where people are employed including those covered by other fire-specific fire-safety legislation (such as those for which a Fire Certificate is in force, or has been applied for under the *Fire Precautions Act 1971*).

In the majority of cases the employer must undertake a fire risk assessment of the workplace to establish what precautions are necessary to ensure that employees are safe in the event of fire. This might include, for example:

• Making sure that the workplace is equipped with fire-fighting equipment, fire detectors and alarms.

- Ensuring that any non-automatic fire-fighting equipment is easily accessible, simple to use and indicated by signs.
- Nominating properly trained employees to implement the necessary fire-safety measures.
- Ensuring that necessary contacts are made with external emergency services, regarding rescue work and fire fighting.
- Ensuring that suitable emergency routes and exits are available.
- Organising a suitable system of maintenance so that all equipment are maintained in efficient working order and in good repair.

Every fire authority has a duty to enforce the Regulations in its area including powers to inspect premises at any time to ensure compliance with the Regulations.

More information on the Fire Precautions (Workplace) Regulations (including a definition of '*Workplace*') is given in Section 5.2.1.

4.3.5 The Fire Precautions Act 1971

The *Fire Precautions Act 1971* is primarily concerned with the provision and maintenance of safe means of escape in case of fire in buildings when they are being used. All buildings put to a designated use under the Act are required to have a fire certificate (issued by the local fire authority) unless they are exempt. Fire certificates are only required for premises which are put to certain 'designated' uses. To date, two designating orders have been made covering:

(1) *hotels or boarding houses* where sleeping accommodation is provided for more than six staff or guests (or some sleeping accommodation is provided above the first floor or below the ground floor)
(2) *factories, offices, shops and railway premises* where more than 20 people are at work at any particular time, or more than 10 people work other than on the ground floor or in factories only where explosive or highly flammable materials are stored or used.

It is possible that an existing building may be put to a designated use where alterations are unnecessary, and which does not constitute a material change of use under the building regulations. Where a fire certificate is needed in these circumstances application must be made to the fire authority for the area in which the building is located (and which is the body which enforces the Act) in the prescribed form in accordance with the requirements of section 5 of the Act. On the other hand, where work is being carried out which must comply with the Building Regulations, the Local Authority or Approved Inspector will deal with the application and carry out the necessary consultations with the fire authority (see Sections 5.3.2 and 5.7.1). There is no need for a separate application to be made to the fire authority.

4.3.6 The Disability Discrimination Act

The purpose of the *Disability Discrimination Act 1995* (DDA) is to end the discrimination faced by disabled people in the areas of employment, the provision of goods, facilities and services, and in the renting or buying of property or land. Therefore, the DDA contains duties to make reasonable adjustments to physical features of premises in certain circumstances.

Duties in the Employment Field
Up to 30 September 2004, section 6 of the DDA sets out the duty of employers with 15 or more employees to make these reasonable adjustments. An exemption from this duty was provided by regulation 8 of the *Disability Discrimination (Employment) Regulations 1996*. Under these regulations an employer was not required to alter any physical characteristic of a building, which was adopted with a view to satisfying the requirements of Part M of the Building Regulations and met those requirements at the time the building works were carried out and continued substantially to meet those requirements. It should be noted that there was no duty on providers of service to the public to make reasonable adjustments to physical features of premises.

From 1 October 2004, the exemption in the 1996 Regulations ceased to apply. Some changes to the duty to make reasonable adjustments were introduced from 1 October 2004 and its coverage was extended to all employers (irrespective of size), a range of other bodies and occupations (e.g. qualifications bodies and partners in business partnerships), and to all those who provide services to the public irrespective of their size.

At present, following the guidance in the current (2004) edition of Approved Document M is not a requirement for satisfying the duties to make reasonable adjustments to the physical features of premises. The Government is considering whether to introduce an exemption so that features constructed or installed in accordance with this edition of Approved Document M will satisfy the duties to make reasonable adjustments and will make an announcement in due course.

4.3.7 The Building (Local Authority Charges) Regulations 1998

These regulations (which do not apply to Approved Inspectors) authorise local authorities to fix and recover charges for the performance of their building control functions according to a scheme governed by principles prescribed in the regulations (see also Sections 5.2.1 and 5.2.2). Therefore, they have the effect of making each local authority responsible for fixing the amount of its own building regulations control charges for the following functions:

- a plan charge (for passing or rejection of deposited plans of proposed work)
- an inspection charge (for inspection of works in progress)

- a building notice charge (where the building notice procedure is adopted; this is the sum of the plan and inspection charges which becomes payable when the notice is given to the local authority; and this may be varied with the consent of the local authority, see Section 5.2.1)
- a reversion charge (where the Approved Inspector system is used and the initial notice is cancelled so that control reverts to the local authority, see Sections 5.8.3 and 6.4)
- a regularisation charge (for unauthorised building work, see regulation 21 in Appendix 1 and Section 6.2.1).

Although local authorities have complete freedom to set the levels of the charges they impose, the income derived from the charges over a continuous 3-year period (commencing on the date on which the local authority fixes its charges) must be not less than the cost directly or indirectly (the 'proper costs') incurred in accordance with proper accounting principles. Local authorities must estimate the aggregate of their proper costs over the 3-year period before fixing their charges and they must issue a statement at the end of each financial year, which sets out the details of the charging scheme and shows the amount of the income and proper costs. The following points should also be noted:

- Local authorities are not permitted to levy charges in relation to the control of work which is solely for the benefit of disabled people.
- A local authority must give at least 7 days notice before bringing its charges into effect, although the charges can be amended whenever necessary provided the due notice is given on each occasion.
- Where the proper costs are less than £450,000 over the 3-year accounting period or where at least 65% of the charges received are connected with work to small domestic buildings, such as extensions, garages and carports, the income derived must cover at least 90% of costs.
- With the agreement of the local authority, charges for work may be paid in instalments.
- Reduced charges may be levied for repetitive building work. This could be part of the same submission (such as the construction of a number of houses on a single estate), or for similar work on different sites submitted by the same person (such as work to terraced dwellings under a refurbishment scheme).

Basis of charging

The following works must be charged by reference to the floor area of the building or extension concerned:

- the erection of small domestic buildings (up to 300 square metres in floor area), or
- small detached garages and carports with floor area up to 40 square metres, or
- domestic extensions (including associated access work) with floor areas up to 60 square metres.

Where it is intended to erect a number of extensions to a building their floor areas must be aggregated.

All other types of work are subject to charges related to an estimate of the cost of the work. Local authorities are permitted to forego the inspection charge if the value of the work is less than £5000, however, a local authority may not charge for building work which comprises:

• the installation of cavity fill material into the cavity wall of a building, or
• the installation of unvented hot water systems.

where this work forms part of other building work and is being carried out in accordance with Parts D and G, respectively, of Schedule 1 to the Building Regulations 2000.

Charges are also payable where application is made to the Secretary of State for determination of questions under sections 16 and 50 of the Building Act 1984 (see Section 6.3.2).

In order to achieve some form of national consistency over levels of charges, the Local Government Association has published a 'model scheme' which local authorities are urged to follow.

The Charges Regulations came into force on 1 April 1999, and the cumulative charges income and costs figures for local building control authorities for the first 3-year accounting period were delivered to the ODPM in 2002. Since that date local authorities have been required to submit annual returns of accrual and surpluses. As a result, it is apparent that around 50% of building control authorities are continuing to accrue what the ODPM and HM Treasury consider to be significant surpluses. Since the Charges Regulations do not permit local authorities to levy charges other than for the purpose of recovering their costs, where they are accruing surplus income this could be seen as a tax rather than a charge, and local authorities have no statutory powers to levy such taxes. As a result, the ODPM expects local authorities with significant surpluses to reduce their charges appropriately to ensure that they more accurately equate costs with income. If this happens, the knock-on effect for building control charges as a whole (including the Approved Inspector system) could be considerable since there has already been a 'price war' between the public and private sectors for some time.

4.3.8 The Construction Products Regulations 1991

These regulations (as amended) apply to any construction products which are produced for incorporation in a permanent manner in construction works, and which were supplied after 27 December 1991. The regulations are designed to ensure that when products are used in construction work, the work itself will satisfy any of the following six relevant 'essential requirements' of the Construction Products Directive:

(1) Mechanical resistance and stability.
(2) Safety in case of fire.

(3) Hygiene, health and the environment.
(4) Safety in use.
(5) Protection against noise.
(6) Energy economy and heat retention.

The CE Mark

One way of showing that a product complies with the essential requirements is by means of the CE Mark. For the purpose of issuing CE Marks, each Member State of the European Union is required to maintain a register of designated or notified bodies which identifies:

- Test laboratories (for testing samples).
- Inspection bodies (for inspecting factories and processes).
- Certification authorities (to interpret results).

Satisfactory appraisal by the above-mentioned bodies would allow a manufacturer to affix the CE Mark to his product. The CE Mark is not a quality mark. It signifies only that the product satisfies the essential requirements and may legally be placed on the market. It is primarily intended for enforcement officers, however, Approved Document to regulation 7 makes it clear that materials bearing the CE marking must be accepted as complying with regulation 7 (*Materials and workmanship*) of the building regulations if they are appropriate for the circumstances in which they are used. The Approved Document qualifies this by saying that a CE marked product can only be rejected by a building control body on the basis that:

(a) its performance is not in accordance with its technical specification, or
(b) where a particular declared value or class of performance is stated for a product, the resultant value does not meet Building Regulation requirements.

The burden of proof is on the controlling body which is obliged to notify the Trading Standards Officer so that the UK Government can notify the Commission.

4.3.9 The Gas Safety (Installation and Use) Regulations 1998

The *Gas Safety (Installation and Use) Regulations 1998* are intended to control the risks associated with the use of gas supplied from gas storage vessels or via mains pipes. They do this by requiring that:

- work on gas installations is carried out by a competent person
- the competent person's employer is recognised as being approved by the Health and Safety Executive (i.e. a member of the Council for Registered Gas Installers (CORGI))
- gas appliances, gas fittings, installation pipework and flues are installed safely (including the carrying out of checks to ensure compliance with the Regulations)

- any flue is installed in a safe position
- alterations carried out to any premises in which a gas fitting or gas storage vessel is fitted are expedited without adversely affecting safety or causing non-compliance with the Regulations
- gas appliances and installation pipework are maintained in a safe condition in workplaces covered by the Regulations
- restrictions are placed on the installation of certain types of gas appliances in bathrooms and sleeping accommodation, or in cupboards or compartments in such rooms
- all gas appliances and pipework installed in rented premises are maintained in a safe condition (which includes the keeping of maintenance records)
- liquefied petroleum gas (LPG) storage vessels, and LPG-fired appliances fitted with automatic ignition devices or pilot lights are not installed in cellars or basements.

The Building Regulations are directly linked with the Gas Safety (Installation and Use Regulations), since regulation 12(5) and Schedule 2A permit the installation of a heat-producing gas appliance by '*a person, or an employee of a person, who is a member of a class of persons approved in accordance with regulation 3 of the Gas Safety (Installation and Use) Regulations 1998*' (i.e. a member of CORGI) without having to submit a building notice or deposit full plans (see Section 5.2.1).

4.3.10 The London Building Acts 1930 to 1982

When the first National Building Regulations came into force (1 February 1966), they did not apply to Inner London, which continued to be dealt with by the Greater London Council under the London Building Acts 1930 to 1982 and the building byelaws made under them. Enforcement was carried out by District Surveyors who worked for the Inner London Boroughs and the City of London.

This changed on 6 January 1986 when the Building (Inner London) Regulations 1985 came into force. Following the abolition of the Greater London Council on 1 April 1986, its building control functions and those of district surveyors under the London Building Acts were transferred to the Common Council of the City of London and the 12 Inner London Boroughs (Westminster, Camden, Islington, Hackney, Tower Hamlets, Greenwich, Lewisham, Southwark, Lambeth, Wandsworth, Hammersmith and Fulham, and Kensington and Chelsea).

As a result of the new regulations, Inner London building control procedures became essentially the same as elsewhere in England and Wales, since all the London Building Byelaws and many sections of the London Building Acts were repealed. Certain sections of the London Building Acts, which were deemed to be peculiar to London, were retained with revisions and these remain in force. Many of the powers are discretionary and were applied with a fair degree of uniformity whilst under the control of the former Greater London Council. Unfortunately, each relevant local authority now uses the powers within its own procedures and policies, and this has led to a significant degree of variance in interpretation between the authorities.

The most important provisions of the London Building Acts which were retained are listed below.

Dangerous and noxious businesses

Sections 143 and 144 of the *London Building Act 1930* prohibit the erection of any building within 50 feet of any building used for a dangerous and/or noxious business (see section 5 of the *London Building Act 1930* for a definition of dangerous business and section 4 of the *London Building Acts (Amendment) Act 1939* for a definition of noxious business). It is also not permitted to establish or carry on any such business in a building within 40 feet of any public way or 50 feet from any other building. The requirements of these sections may be waived by the local authority after consultation with the fire authority.

Special fire precautions in high and certain large buildings

Section 20 of the *London Building Acts (Amendment) Act 1939* applies special fire precautions to the proposed high or large buildings; that is, a building which has a storey at a greater height than 30 metres (or 25 metres if the area of the building exceeds 930 square metres), or is a warehouse or building used for trade or manufacture with a cubic capacity over 7100 cubic metres, by requiring that they be divided up by division walls as defined in the Act so that no division exceeds 7100 cubic metres. Any openings in the division walls must be protected by doors or shutters in accordance with section 21 of the Act (see below).

Section 20 is activated when any notice is given to the local authority in Inner London in respect of building work to be carried out under the building regulations. In this respect, the procedure is similar to that for the linked powers (see Section 4.3.1) and applies equally to both the Local Authority and Approved Inspector control systems. The owner or occupier of the building must deposit with the local authority two copies of the plans and other relevant information in accordance with local authority regulations specifying the level of detail required. These regulations may be made by the local authority under section 145(5) of the *London Building Acts (Amendment) Act 1939* or they may be the original regulations made by the Greater London Council on 2 July 1974.

After consulting the London Fire and Emergency Planning Authority the local authority may impose conditions for the provision and maintenance of:

- Fire alarms and automatic fire detection systems.
- Fire extinguishing installations and appliances.
- Effective means of removing smoke in the event of fire.
- Adequate means of access to the site of the building, and to the interior and exterior for fire brigade personnel and appliances.

Additionally, there are extra requirements for areas of 'special fire risks' such as boiler rooms.

The conditions which can be imposed are not prescribed by statute but are based on guidance contained in the London District Surveyors Association Fire Safety Guide No. 1 which is currently undergoing revision to remove anomalies and unnecessary conflict with the building regulations.

Since the advent of the Approved Inspector control system, section 20 has come in for a considerable amount of criticism. In general, it is seen by all parties (except the local authorities) as an unnecessary piece of extra 'red tape' that duplicates much of what is in the current building regulations. Under section 3 of the *GLC General Powers Act 1982* local authorities in Inner London are permitted to charge fees for processing section 20 applications and these vary greatly from one authority to another. In some circles it is also regarded as anti-competitive since it can be administered only by local authorities and can be seen to give them an unfair advantage over Approved Inspectors in cases where the client wants all approvals to be obtained from one source. Additionally, some Approved Inspectors have claimed that where they have been employed to carry out the building control function it is often the case that the local authority will impose additional requirements on the building owner, which go considerably beyond those of the building regulations and result in unnecessary extra expense. Even interpretation of the guidance contained in the LDA Fire Safety Guide varies considerably from one local authority to another. In order to counter these problems some of the larger Approved Inspectors now employ ex-Inner London local authority building control surveyors who have had expertise in administering section 20 for their former employers. These Approved Inspectors offer clients a section 20 consultancy service which aims to foresee any difficulties which might arise when the section 20 application is submitted to the local authority. The consultant will also negotiate with the local authority on behalf of the client over the conditions which may be imposed by the local authority.

Uniting of buildings

Section 21 of the *London Building Acts (Amendment) Act 1939* states that local authority consent is required if two buildings are united by making an opening in a party wall or external wall separating buildings or when buildings are connected by access without passing into the external air. This does not apply if the buildings are in one ownership and if united would comply with the London Building Acts.

Special and temporary buildings and structures

Sections 29 to 31 of the *London Building Acts (Amendment) Act 1939* deal with the erection and retention of certain temporary buildings which need the consent of the local authority and are not covered by other legislation. In practice, this section is used to control a range of structures which are neither controlled nor exempt under the building regulations, such as radio or lighting masts, freestanding walls and fences exceeding 1.83 metres in height, advertising hoardings, etc. Consent can be time limited and subject to renewal if the structure is subject to deterioration or can be given outright without limit. Fees are charged for the issue and renewal of consents, and these vary between local authorities.

Means of escape in old buildings

Sections 35 to 37 of the *London Building Acts (Amendment) Act 1939* enable local authorities to require the provision of means of escape in certain old buildings constructed before 1939. In most cases control over means of escape in the categories of buildings described in section 35 has been superseded by much later legislation which is more effective. Nevertheless, some local authorities still use these powers, particularly in respect of old flat conversions which have never received any kind of approval.

Maintenance of means of escape

Section 133 of the *London Building Acts (Amendment) Act 1939* requires that all means of escape provisions imposed under sections 33 and 34 of the same Act be maintained in good working condition.

Dangerous and neglected structures

Part VII of the *London Building Acts (Amendment) Act 1939* enables local authorities to deal with dangerous buildings and structures by serving a notice on the owner requiring them to take down, repair or otherwise secure the parts certified as dangerous. Non-compliance with the notice can result in the local authority applying for a court order, and where this is not complied with the local authority may carry out the work and recover costs from the owner. Immediately dangerous buildings and structures can be dealt with by the local authority without the service of a notice. This part of the Act was applied to all Outer London Boroughs (except Barnet) by virtue of the *London Local Authorities Act 2000*. The powers to deal with dangerous structures in all other parts of England and Wales are contained in sections 77 and 78 of the Building Act 1984 (see Section 4.3.1).

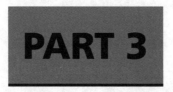

PART 3

Administration

5

Control systems

USING A LOCAL AUTHORITY

5.1 Background

Reference to Chapter 2 will reveal that, for practical purposes (and with the exception of Inner London), the origin of the building control system for England and Wales lies in the late Victorian era with the passing of the Public Health Act 1875. This Act empowered local authorities to make and enforce byelaws (i.e. local laws) for their districts aimed at securing the health and safety of the public when using or being in close proximity to buildings.

The Act allowed byelaws to be made covering the sufficiency of air space about buildings, ventilation, drainage, sanitation, and basic constructional requirements covering freedom from dampness, structural stability, etc. National model byelaws were published by the Local Government Board (a department of state to supervise local authorities), their adoption being voluntary, and most local authorities embraced the model byelaws whilst adding their own local variations. Since the byelaws were administered by more than 1400 local authorities, considerable variations existed over relatively small regional areas.

Adjustments to the system brought about by the *Public Health Act 1936* did little to promote uniformity of control. Eventually, this was dealt with by the *Public Health Act 1961*, which gave the Secretary of State the powers to make National Building Regulations for England and Wales (although Scotland, Northern Ireland and Inner London were not included in the reforms). The first National Regulations for England and Wales (the *Building Regulations 1965*) came into force on 1 February 1966.

The Regulations passed through numerous amendments and consolidations and eventually became the *Building Regulations 1976* (at the same time increasing in size from 169 to 306 pages with corresponding increases in detail and complexity).

In 1974, the reorganisation of local government boundaries (brought about by the *Local Government Act 1972*) reduced the number of local authorities to about 400. In many cases reorganisation created co-ordination and control problems for the new expanded local authorities and led to an increase in staff numbers and bureaucracy.

Although the building control system was seen to produce safe buildings in which fire and serious structural failure were rare, during the 1970s persistent criticisms existed on other counts. In particular, the system was perceived to be more cumbersome and bureaucratic than necessary. Additionally, the detailed form of the Regulations was blamed for inflexibility in use, and was said to inhibit innovation and impose unnecessary costs.

In February 1981 the Secretary of State for the Environment issued a command paper, *The Future of Building Control in England and Wales*. This paper contained a range of proposals dealing with revision of the Building Regulations and with the processes of control. It included detailed proposals which would allow private 'certifiers' (later to be called 'Approved Inspectors') to check plans and inspect work on site, and provided for the recasting of the Building Regulations as a minimum number of functional requirements, with supporting approved Codes of Practice (later to be called 'Approved Documents'). On the basis of this command paper the Department of the Environment sets about the task of completely recasting the Building Regulations and of producing a new Act of Parliament which would amend the law relating to the structure of the Building Regulations and the supervision of work subject to Building Regulation control.

Eventually, these proposals saw the light of day as the *Building Act 1984*, which came into force on 1 December 1984. This Act consolidated various building control statutes enacted over the previous 90 years and helped to ease considerably the task of those engaged in building control. The *Building Act 1984* (as amended) is the principal controlling legislation for both the Local Authority and Approved Inspector building control processes.

5.2 Procedures

Reference to Flow chart 5.1 will show the outlines of the two control procedures that may be used by a person intending to carry out any building work to which the regulations apply. Although the chart indicates that two procedures are available (Building Notice or Full Plans) the passing of the *Fire Precautions (Workplace) Regulations 1997* has meant that use of the building notice is now mainly restricted to work on dwellings.

5.2.1 The building notice

The ability to serve a building notice on a local authority (as an alternative to the deposit of detailed plans of a proposal) was introduced in 1985 as a result of the passing of the *Building Act 1984*. The idea, which was copied from the former Inner London control system, does not involve any formal approval or passing of plans by the local authority.

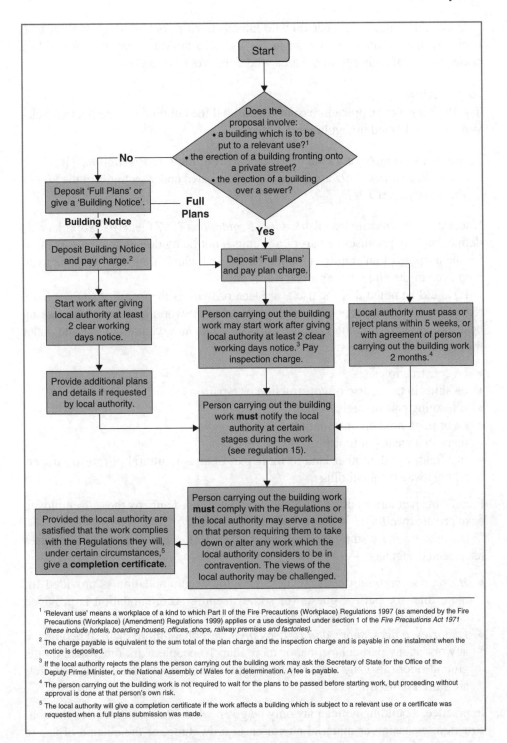

Flow chart 5.1 The local authority system of building control.

Content within the flow chart:

Start

Does the proposal involve:
• a building which is to be put to a relevant use?[1]
• the erection of a building fronting onto a private street?
• the erection of a building over a sewer?

No

Deposit 'Full Plans' or give a 'Building Notice'.

Building Notice

Full Plans

Yes

Deposit Building Notice and pay charge.[2]

Deposit 'Full Plans' and pay plan charge.

Start work after giving local authority at least 2 clear working days notice.

Person carrying out the building work may start work after giving local authority at least 2 clear working days notice.[3] Pay inspection charge.

Local authority must pass or reject plans within 5 weeks, or with agreement of person carrying out the building work 2 months.[4]

Provide additional plans and details if requested by local authority.

Person carrying out the building work **must** notify the local authority at certain stages during the work (see regulation 15).

Provided the local authority are satisfied that the work complies with the Regulations they will, under certain circumstances,[5] give a **completion certificate**.

Person carrying out the building work **must** comply with the Regulations or the local authority may serve a notice on that person requiring them to take down or alter any work which the local authority considers to be in contravention. The views of the local authority may be challenged.

[1] 'Relevant use' means a workplace of a kind to which Part II of the Fire Precautions (Workplace) Regulations 1997 (as amended by the Fire Precautions (Workplace) (Amendment) Regulations 1999) applies or a use designated under section 1 of the *Fire Precautions Act 1971* *(these include hotels, boarding houses, offices, shops, railway premises and factories).*

[2] The charge payable is equivalent to the sum total of the plan charge and the inspection charge and is payable in one instalment when the notice is deposited.

[3] If the local authority rejects the plans the person carrying out the building work may ask the Secretary of State for the Office of the Deputy Prime Minister, or the National Assembly of Wales for a determination. A fee is payable.

[4] The person carrying out the building work is not required to wait for the plans to be passed before starting work, but proceeding without approval is done at that person's own risk.

[5] The local authority will give a completion certificate if the work affects a building which is subject to a relevant use or a certificate was requested when a full plans submission was made.

Under this procedure, a person who intends to carry out building work or who wants to make a material change in the use of a building may give a building notice to a local authority in the area where the work is situated.

Application

The Building Notice procedure cannot be used if the building is to be put to a 'relevant use' as defined in regulation 12; that is,

> use as a workplace of a kind to which Part II of the *Fire Precautions (Workplace) Regulations 1997* applies or a use designated under section 1 of the *Fire Precautions Act 1971*.

Under the *Fire Precautions (Workplace) Regulations 1997* the term 'workplace' is defined as 'any premises or part of a premise, not being domestic premises, used for the purpose of an employer's undertaking and which are made available to an employee of the employer as a place of work'.

It should be noted that the use as a place of work is the important factor here, regardless of whether or not the employer is using the workplace as a profit-making business. However, certain types of workplaces are an exception to the rule under the regulations as follows:

- on construction sites
- on ships in the course of construction or repair
- if forming part of a mine facility
- on or in an offshore installation
- on or in a means of transport
- in a field, wood or other land forming part of an agricultural or forestry undertaking away from buildings.

Finally, the regulations do not apply to workplaces used only by the self-employed or to private dwellings.

Under *section 1* of the *Fire Precautions Act 1971*, the following building uses have been designated:

- *Hotels and boarding houses* where sleeping accommodation is provided for more than six staff or guests (or some sleeping accommodation is provided above the first floor or below the ground floor).
- *Factories, offices, shops and railway premises* where more than 20 people are at work at any particular time, or more than 10 people work other than on the ground floor (also any factory where explosive or highly flammable materials are stored or used).

In practice, a building notice may only be given where the work involves the construction, extension or material alteration of a dwelling, or the change of use to provide one or more dwellings. This option may be further reduced if:

- the building (including a dwelling) fronts on to a private street or

- the work involves construction over or adjacent to a drain, sewer or disposal main that is shown on the sewerage records of the sewerage undertaker.

Form of notice

There is no official form of building notice; however, it must be signed by, or on behalf of, the person intending to carry out the work and must contain or be accompanied by the following information:

- The name and address of the person who intends to carry out the building work.
- A statement that the notice is given in accordance with regulation 12(2)(a).
- A description of the proposed building work or material change of use.
- Particulars of the location of the building to which the proposal relates and the use or intended use of that building.
- For work involving the erection or extension of a building, a plan to a scale of not less than 1:1250 showing:
    - (i) its size and position, including its relationship to adjoining boundaries
    - (ii) the boundaries of its curtilage, and the size, position and use of every other building or proposed building within that curtilage
    - (iii) the width and position of any street on or within the boundaries of its curtilage.
- A statement specifying the number of storeys (counting each basement level as one storey), in the building to which the proposal relates.
- Particulars of the provision to be made for the drainage of the building or extension, and the steps to be taken to comply with any local legislation which applies.

A typical notice is illustrated in Figure 5.1. Full details of the building notice procedure are given in regulations 12 and 13.

Additional details and information

For certain minor works the building notice should be supplemented with additional details as follows:

(1)   The insertion of insulating material into the cavity walls of a building:
   - the name and type of insulating material to be used
   - the name of any European Technical Approval issuing body, for example, British Board of Agrément (BBA), which has approved the insulating material, or details of any national standard of a Member State of the European Economic Area (e.g. British Standard) to which the material conforms
   - information regarding those parts of Schedule 1 of the Regulations under which the approval body (e.g. the BBA) has approved the insulating material (e.g. paragraph C4 deals with resistance to the passage of moisture through the walls of a building and is supported by Approved Document C which contains guidance on the installation of cavity insulation so that moisture penetration is avoided)
   - the name of the body which has approved the installer of the insulating material.

Building notice

The Building Act 1984	The Building Regulations 2000

Building Control Services	**Alford Borough Council** Alford, Hants BH99 9ZZ

This form is to be filled in by the person who intends to carry out building work or his/her agent. If difficulty is experienced in filling in this form please refer to the notes overleaf or consult the office indicated above.

1	**Applicant's details** Name: Address: Postcode: Telephone: Fax:
2	**Agent's details** (if applicable) Name: Address: Postcode: Telephone: Fax:
3	**Location of building to which work relates** Address: Postcode:
4	**Proposed work** (e.g. domestic extension, internal alterations, re-roofing etc.) Description:
5	**Use of building** Domestic ☐ Non-domestic ☐ (tick box) 1. If new building or extension please state proposed use _ _ _ _ _ _ _ _ _ _ _ _ _ _ _ _ _ 2. If existing building please state present use _ A Building Notice cannot be submitted if the building is to be put to a use as a workplace of a kind to which Part II of the *Fire Precautions (Workplace) Regulations 1997* applies or a use designated under section 1 of the *Fire Precautions Act 1971*. Therefore, where the work relates to a hotel, boarding house, shop, office, factory, railway premise or any place of work Full Plans must be submitted.
6	**Charges** 1. If Schedule 1 work please state the total number of dwellings and types – Total No. No. of types 2. If schedule 2 work please state floor area: m^2 3. If schedule 3 work please state the estimated cost of work excluding VAT: £ Building Notice Charge: plus VAT: Total: £ £ £
7	**Additional information:** 1. Number of storeys: 2. Number of bedrooms: 3. Means of water supply:
8	**Statement** This notice is given in relation to the building work as described, is submitted in accordance with Regulation 13 and is accompanied by the appropriate payment. This Building Notice shall cease to have effect three years from the date submitted to the Local Authority unless the work has previously commenced and the appropriate notification given to the Council. Name: Signature: Date:

Figure 5.1 Typical building notice.

(2) The provision of a hot water storage system covered by paragraph G3 of
     Schedule 1:
     – the name, make, model and type of system
     – the name of the body which has approved or certified that the system is
       capable of satisfying the requirements of G3
     – the name of the body which has issued any current registered operative
       identity card to the installer of the system.

The building notice procedure is designed to make it easier for people to apply for
regulation approval for relatively minor work where detailed building plans are
not really necessary (e.g. a structural alteration involving the removal of a wall to
make a through lounge in a dwelling, or the conversion of a spare bedroom to provide an additional bathroom). However, there may be circumstances where additional information will need to be provided so that the local authority can ensure
that the regulations have been complied with. In the first example given above it
may be that structural calculations are needed to confirm that the structure above
the removed wall has been properly supported.

   Therefore, the local authority is entitled to ask for additional plans and information. The local authority must specify in writing the information that is required
and a time limit can be laid down for its provision. Curiously, although the local
authority can demand the supply of additional material this is not deemed to have
been 'deposited' under the regulations therefore, there are no powers to pass or
reject it. Clearly, the extent of the additional information that is requested will
depend on the type and complexity of the work being carried out. For example, the
construction of a new dwelling-house using the building notice procedure would
undoubtedly result in a considerable amount of additional information being
sought by the local authority. Since the developer, having produced this information, would not have the benefit of having the plans passed by the local authority,
it is probably more sensible to use the full plans procedure (see Section 5.2.2) in
such cases. As is indicated above, the Building Notice procedure is the most useful for minor alterations and extensions to dwellings, where the production of
scale plans is not usually needed.

Duration
A building notice automatically ceases to have effect for 3 years from the date on
which it was given to the local authority, if the work has not started or the change
of use has not taken place. This contrasts with the situation regarding the deposit
of full plans (see Section 5.2.2).

Charges
Local authorities are empowered under the *Building (Local Authority Charges)
Regulations 1998* to fix and to recover charges for the performance of their building regulation control functions in line with principles prescribed in the 'Charges'
regulations. Usually, the charge for a building notice must be paid when the notice

is given to the local authority, or the notice is not deemed to have been given. However, the local authority may agree to the charge being paid in instalments, although in this case it will lay down the amounts to be paid and the dates on which the sums will be due.

Advantages and disadvantages

From a practical viewpoint, the main advantage of using a building notice lies in the fact that work can be started as soon as the notice has been given to the local authority and the required charge has been paid, although there is still a requirement, which also applies in the case of the full plans procedure, that the local authority be given at least 2 clear working days notice before the work starts. The main disadvantages are that there is no procedure for approval of information supplied, the local authority is not required to acknowledge receipt of the notice (although most do) and the local authority is not obliged to issue a completion certificate when the work has been finished. Therefore, on resale of the property at a later date it may be difficult to prove to a prospective purchaser that building regulation compliance was sought for alterations, etc. and this may delay the sale.

5.2.2 Full plans

This is the 'traditional' route for ensuring compliance with the regulations whereby full plans and supporting information are deposited with the relevant local authority in accordance with section 16 of the *Building Act 1984* as supplemented by regulation 14. It is important to note that the term 'full plans' includes a great deal more than just drawings. The *Building Act 1984, section 126* defines 'plans' as including drawings of any description and specifications, or other information in any form.

The 'linked powers'

Under section 16 the local authority must pass the deposited plans unless they are defective or show a contravention of the regulations. In the past, deposit of plans for the purposes of building regulation compliance triggered off a number of so-called 'linked powers' (other sections of the Building Act which came into effect on deposit of plans under the regulations) and the plans could be rejected if they did not show compliance with other provisions.

In the 20 years since the Act came into force most of these linked powers have been replaced by changes in the regulations themselves. The only remaining section relates to the provision for the occupants of a house of a 'supply of wholesome water sufficient for their domestic purposes'. This may be found in section 25 of the *Building Act 1984*. The water supply can be provided in any of the following ways:

- by connecting the house to a water supply provided by a water undertaker (i.e. a mains supply)

- where it is not reasonable to connect to a mains supply (in remote country districts there may be no mains supply) by taking the water into the house by means of a pipe (e.g. from a well or spring)
- where circumstances exist which make either of the foregoing solutions unreasonable, the supply of water may be located within a reasonable distance of the house.

This last solution is interesting when considered against the requirements of paragraph G2 of Schedule 1 to the Building Regulations. G2 demands that in a dwelling a '*bathroom shall be provided containing either a fixed bath or shower bath, and there shall be a suitable installation for the provision of hot and cold water to the bath or shower bath*'. It is difficult to see how this could be achieved if a water supply is not provided in the dwelling.

The wholesomeness of water is judged by reference to section 67 of the *Water Industry Act 1991 (standards of wholesomeness of water)* as read with regulations made under that Act (i.e. the *Water Supply (Water Quality) Regulations 2001*).

Approved persons
Section 16 also allows (in subsection (9)) local authorities to deal with plans accompanied by a certificate from an 'approved person'. The term 'approved person' under section 16(9) of the 1984 Act should not be confused with approved inspectors under sections 47 to 54. Where plans are accompanied by such a certificate from an approved person that the proposed work will comply with such provisions of the regulations as may be prescribed, and evidence of satisfactory insurance cover is provided, the local authority may not reject the plans on the grounds that they are defective, or take proceedings under section 35 of the 1984 Act for contravention. So far, regulations 4 and 6 have been prescribed insofar as they relate to Part A (Structure) and Part L (*Conservation of fuel and power*). To date, no approved persons have been so designated to issue certificates (see also Section 3.2.4).

Time periods and charges
Once the required information has been deposited with the local authority it has 5 weeks in which to pass or reject the plans. This period can be extended up to 2 calendar-months from the date of deposit provided that the person carrying out the building work gives written consent to the extension of time. This written consent must be given before the 5-week period expires. These time periods will only commence on deposit of plans if the applicant has paid a plan charge at the same time as the plans are deposited (which may also involve the submission of a 'reasonable estimate' of the cost of the works in certain circumstances). The powers of local authorities to levy charges are contained in the *Building (Local Authority Charges) Regulations 1998* whereby they are able to fix and recover charges for the performance of their building regulation control functions in line with principles prescribed in the 'Charges' Regulations. By the same token if the local authority does not give a decision within the statutory time periods it must refund the plan charge.

Duties of a local authority to pass or reject plans

The local authority must pass the plans of any proposed work deposited in accordance with the regulations unless the plans are defective or show that the proposed work would contravene the regulations. The term 'defective' is not defined in either the regulations or the Building Act but is usually taken to mean that insufficient or contradictory information has been supplied. If the plans cannot be passed as they stand the local authority must either:

- issue a notice of rejection detailing those parts of the Regulations or Building Act which have not been complied with, or otherwise why the plans are considered to be defective or
- pass the plans subject to conditions.

Once the plans have been passed by the local authority, the work must be commenced within a period of 3 years from the date of deposit or the approval will lapse. The onus is on the local authority to give formal notice to the applicant to this effect.

Conditional passing of plans

Plans can only be passed conditionally if the applicant has issued a written request to the local authority giving consent for them to do this. The local authority may only apply the following conditions when taking this route:

- that certain specified modifications be made to the deposited plans
- that such further plans as they may specify be deposited.

The consent to allow the local authority to pass the plans conditionally is usually facilitated by means of a question to that effect included in the application form. Where the local authority specifies the additional plans that are to be deposited it is implied that these should be submitted before the work contained in them, is carried out since the plans referred to are of 'proposed' work. However, this would not appear to allow the local authority to insist on an actual timetable for deposit of the plans.

In practice, the facility to submit further plans as the job progresses is the only practical way of dealing with large and complex projects which may take several years to complete. In many cases different sections of the work are put out to different specialists to be designed as packages and all the detailed design information may not be available at the start of the project. However, it does depend on close co-operation between the contractor, design specialist and local authority building control team, if problems and costly mistakes are to be avoided.

Making a full plans submission

Regulation 14 specifies that plans must be deposited in duplicate with the local authority being authorised to retain one set. It is usual for the other set to be returned to the applicant once a decision has been made although the local authority is not required by law to do this. Two additional sets of plans must be submitted where

Part B (Fire Safety) of Schedule 1 imposes a requirement in relation to the work and the local authority may retain both the additional sets. These extra plans are required so that the local authority can carry out its consultations with the fire authority. Additional sets of plans are not required where the proposed work relates to the erection, extension or material alteration of a dwelling-house or flat.

A full plans submission must contain the following information:

- A description of the proposed building work or material change of use, and the plans, particulars and statements required for a building notice (see Form of notice under Section 5.2.1).
- Details of the precautions to be taken where the work relates to the erection or extension of a building, or to works of underpinning, adjacent to or over a drain, sewer or disposal main that is shown on the sewerage records of the sewerage undertaker (i.e. in cases where compliance is required with paragraph H4 of Schedule 1).
- Any other plans and information, etc. that is needed to show that the work would comply with the regulations.

The additional information will often be in the form of calculations, for example, to prove structural stability or thermal insulation values. It may also be necessary in certain circumstances to provide details of tests for airtightness, continuity of thermal insulation or sound insulation.

The full plans submission will need to be accompanied by a statement as to whether the building is put or is intended to be put to a relevant use (see Application, for definition under Section 5.2.1).

When making a full plans submission, it is possible for the person carrying out building work to request that on completion of the work, the person wishes the local authority to issue a completion certificate. More details of this are included below.

There is no official full plans application form, although each local authority will be able to supply a potential applicant with its own personalised form. There is no legal requirement to use such a form and most local authority forms vary in detail; however, all should contain at least the information shown in Figure 5.2. The notes contained above will assist in filling in the typical form shown in Figure 5.2 or any standard local authority submission form.

5.3 Consultations

Once a full plans submission has been deposited with a local authority and the necessary charge has been received, it will carry out a thorough check of the information provided in order to establish the extent to which the proposals comply with the regulations. In many cases this checking process will involve the local authority entering into consultation with other statutory bodies. It should be noted that the building notice procedure cannot be used in cases where the following consultations become necessary.

Full plans submission

The Building Act 1984	The Building Regulations

Building Control Services	**Alford Borough Council** Alford, Hants BH99 9ZZ

This form is to be filled in by the person who intends to carry out building work or his/her agent. If difficulty is experienced in filling in this form please refer to the notes overleaf or consult the office indicated above.

1	**Applicant's details** Name: Address: Postcode: Telephone: Fax:
2	**Agent's details** (if applicable) Name: Address: Postcode: Telephone: Fax:
3	**Location of building to which work relates** Address: Postcode:
4	**Proposed work** (e.g. domestic extension, internal alterations, re-roofing etc.) Description:
5	**Use of building** Domestic ☐ Non-domestic ☐ (tick box) 1. If new building or extension please state proposed use _ _ _ _ _ _ _ _ _ _ _ _ _ _ _ _ _ _ _ 2. If existing building please state present use _
6	**Building Regulation Charges Details** 1. If Schedule 1 work (new dwellings) please state the total number of dwellings and types Total No. No. of types 2. If schedule 2 work (domestic extension) please state floor area: m² 3. If schedule 3 work (Other Work) please state the estimated cost of work excluding VAT: £ Plan Charge: £ plus VAT: £ Total: £ Inspection Charge: £ plus VAT: £ Total: £

7	**Additional information**
	1. Number of storeys: 2. Means of water supply:
	3. Method of drainage (a) Foul water:
	                        (b) Surface water:
	4. Is there an intention to build over a drain, sewer or disposal main? Yes ☐ No ☐
	   If yes please provide details of how it is intended to comply with paragraph H4 of
	   Schedule 1 to the Building Regulations

8	**Planning Consent** Have you received Planning Consent for this work? Yes ☐ No ☐ If yes, please give Consent No. and date of decision:
9	**Fire Safety Requirements** Is the building currently put to a use, or proposed to be put to a use as a workplace of a kind to which Part II of the *Fire Precautions (Workplace) Regulations 1997* applies or a use designated under section 1 of the *Fire Precautions Act 1971*? Yes ☐ No ☐ Two additional copies of the plans will need to be provided if you have answered yes to the above.

Figure 5.2 Typical full plans submission form.

10	**Conditions** Do you consent to the plans being passed subject to conditions where appropriate: Yes ☐ No ☐
11	**Extension of time** Do you consent to the prescribed time in which the local authority must give a decision being extended to two months: Yes ☐ No ☐
12	**Completion Certificate** Do you wish to be provided with a completion certificate on completion of the works in accordance with regulation 17? Yes ☐ No ☐
13	**Statement** These plans are deposited for the purpose of regulation 12(2)(b) and are accompanied by the appropriate fee. Name: Signature: Date:

Figure 5.2 (Continued).

5.3.1 Sewerage undertaker

Regulation 14A requires that the local authority, consult the sewerage undertaker in cases where it is established that paragraph H4 of Schedule 1 imposes requirements. This relates to the erection or extension of a building, or to works of underpinning, adjacent to or over a drain, sewer or disposal main that is shown on the sewerage records of the sewerage undertaker. Consultations must take place:

- as soon as practicable after the plans have been deposited
- before the local authority issues any completion certificate in relation to the building work.

Additionally, the local authority must:

(a) give the sewerage undertaker sufficient plans to show that the work would comply with the applicable requirements of paragraph H4 of Schedule 1
(b) have regard to any views expressed by the sewerage undertaker
(c) allow the sewerage undertaker 15 days in which to express its views before passing the plans or issuing a completion certificate, unless the sewerage undertaker has expressed its views to them before the expiry of that period.

The consultation procedure is necessary to ensure that the sewerage authority's legal interests (involving access to and structural stability of sewers) are preserved. Further consultation may also be necessary if the actual construction work reveals considerable variation from that shown on the plans.

5.3.2 Fire authority

The other main area where consultation is necessary is in relation to fire safety. In the past, there has been confusion over the relative roles of the two main bodies (the building control body and the fire authority) responsible for control in this area, since the building control body is responsible for proposed building work and the fire authority is responsible for the safety of people in relation to the operation and use of certain buildings once occupied. In order to avoid this confusion (and to prevent the situation arising whereby the fire authority might require additional work to be done to a newly completed building before it can be occupied) the Government issued a guidance document in February 2001 entitled *Building Regulations and Fire Safety Procedural Guidance*. This document makes it clear that for work which involves a building regulations submission, the local authority should take a co-ordinating role with fire authorities and that any recommendations or advice from the fire authority should be channelled through the building control body to the applicant.

It is only when the building is occupied that the co-ordinating role switches to the fire authority, since they are responsible for enforcing general fire-safety matters under the *Fire Precautions Act 1971* and the *Fire Precautions (Workplace) Regulations 1997*. In circumstances where the fire authority may require actual building work to be done in order to comply with the above legislation, they would need to consult the building control body prior to serving an 'enforcement notice' which would oblige a person to make an alteration to a building. Such consultation procures are laid down by statute in section 17 of the *Fire Precautions Act 1971* and regulation 13(5) of the *Fire Precautions (Workplace) Regulations 1997*.

5.4 Control over work in progress

5.4.1 Introduction

Two methods of notifying the local authority have been outlined above, when there is an intention to carry out building work that is covered by the regulations. Irrespective of the method of notification used, local authorities have certain powers to inspect works in progress and are entitled to enter premises in order to enforce the building regulations, or any provisions of the *Building Act 1984* (see *section 95, Power to enter premises*). In order to facilitate this, the onus is placed on the 'person carrying out the building work' to notify the local authority at certain stages in the construction process. The term 'person carrying out building work' is not defined in the regulations; however, decided case law would indicate that this can mean the owner of a building who authorises a contractor to carry out building works on his behalf. Additionally, the *Building Act, section 36* enables the local authority to take action against the owner of the building in cases where the building regulations have been contravened. Enforcement action is considered in detail in Chapter 6.

More usually, the contractor will take responsibility for notifying the local authority, since it is he or she who will have detailed knowledge of the construction programme. It should be noted that whilst the person carrying out the building work is contravening the regulations if he does not give the required notice, there is no equivalent legal requirement for the local authority to carry out an inspection after being notified. To counter this the *Building Control Performance Standards* set out recommendations for the level of service in respect of site inspections for all building control bodies. Local authorities that subscribe to the performance standards will have regard to this when setting out their inspection policies.

5.4.2 Notification methods

The stages of notification are laid down in regulation 15. The means of giving the notices is no longer specified in the regulations; therefore it is up to the person carrying out the building work to agree with the local authority the method to be used. Most local authorities still issue pre-printed postcards with multi-choice options that can be used for any of the 'statutory' notifications that are described below. This is harking back to the days when the notices had to be in writing; however, there is no reason why notification cannot be in the form of an e-mail, text message, phone call, fax or letter. The important thing to remember is that whichever means of notification is used it may be necessary at some later date to provide evidence of the date of service of the notice, especially where work has been covered up before being inspected. All the notice periods are specified in terms of days, and 'day' means any period of 24 hours commencing at midnight and excludes any Saturday, Sunday, Bank holiday or public holiday.

5.4.3 Notice of commencement

Where a person is using the building notice or full plans procedure, and that person proposes to commence controlled building work, they must then notify the local authority in sufficient time so that at least 2 days have elapsed since the end of the day on which he or she gave the notice.

This will enable the local authority to check whether they have already been notified of the proposal either by means of a building notice or a full plans submission.

If a full plans submission has been made, there is no legal necessity to wait for it to be passed before commencing work; however, any work done will be at the risk of the person carrying out the work and if subsequently, the local authority finds that the plans are defective they would be within their rights to ask for any contravening work to be altered or removed so that it did comply. Most local authorities will inspect work before the plans are passed but they will always state that the inspection is done without prejudice to their rights to take action if the plans are later found to be defective or show a contravention of the regulations.

5.4.4 Notice of works ready for inspection

Covering up of the following work must not commence until the due notice has been given to the local authority and at least 1 day has elapsed since the end of the day on which the notice was given:

- any excavation for a foundation
- any foundation
- any damp-proof course
- any concrete or other material laid over a site
- any drain or sewer to which the regulations apply.

The point of these notifications is to give the local authority the opportunity to inspect parts of the construction which become inaccessible once they have been covered up.

Where no notice is given or the work has been covered up before the notice period has elapsed, the local authority has certain powers to require the exposure of the offending work and to have it corrected if it is not in compliance. Full details of the enforcement actions open to local authorities are given in Chapter 6.

5.4.5 Notice to be given after completion of certain stages of work

In the following cases the local authority must be notified within 5 days of the work being completed:

- when a drain or sewer which is subject to the requirements of Part H (*Drainage and waste disposal*) has been laid, haunched or covered up
- the building work which is the subject of the building notice or full plans submission.

In the case of drains, the local authority is entitled to carry out watertightness tests (see regulation 18, *Testing of building work*) and it may be prudent on the part of the contractor to get the drains tested immediately after backfilling of trenches (an operation notorious for causing damage to drains) and before concrete paving or roadway constructions have been placed. From experience, this notice is often overlooked and can result in expensive remedial works involving the reinstatement of hard landscaping if local authority drain tests are left to be done at the end of the job.

The so-called '*notice of completion*' is also often given outside the specified time period (either too early or too late) and it may be quite difficult to know exactly when the works which are the subject of a building regulation submission are, in fact, complete. This is because the contractor will be involved with a complete contract which will often include extensive internal decoration, engineering services and external landscaping, all of which may not be covered by the regulations. Should the contractor wait until the entire contract is complete or should they try to assess when the controlled works only are complete? There can be no clear-cut answer, as each case will be different. The best course of action is to agree with the local authority the stage at which they consider the controlled works to be complete and act accordingly. This is particularly important where the building is subject to certification under the Fire Precautions Act since the fire authority will also be involved in signing off the building.

5.4.6 Notice of occupation before completion

Where a building is being erected, and that building (or any part of it) is to be occupied before completion, the person carrying out the work must give the local authority at least 5 days notice before the building or any part of it is occupied. This obligation to notify the local authority at least 5 days before occupation where the building is to be occupied *before* completion of the controlled works does not negate the need to also notify them within 5 days of completion of the works.

This requirement allows a building to be occupied in phases and the point of the 'occupation' notice is to enable the local authority (and any other bodies with statutory controls over the building, such as the fire authority) to ensure that the part being occupied is in fact safe. For example, it might be unsatisfactory if a means of escape from an occupied part had to pass through a partially completed part of the same building. It would also be necessary to check that fire alarm systems were working in the occupied part and would be unaffected by works continuing in the unoccupied part.

5.4.7 Completion certificate

The local authority only becomes liable to issue a completion certificate when a notice is served on it under regulation 15 stating that the building work has been completed or that a building has been occupied (either in total or in part) before completion, and when the local authority is satisfied, after having taken all reasonable steps, that the relevant requirements of Schedule 1 specified in the certificate have been satisfied. As has been shown in Section 5.2.2, the full plans application form supplied by the local authority will usually contain a 'tick box' indicating that the applicant wishes the local authority to issue a completion certificate when the works are satisfactorily completed. This certificate will relate to all the applicable requirements of Schedule 1. Where a 'tick box' is not provided the applicant will have to formally request such a certificate. If this is not done the local authority will only be under an obligation to issue a completion certificate for work which is subject to the requirements of Part B of Schedule 1 (*Fire safety*). A completion certificate is a valuable document and will almost certainly be requested by the purchaser's solicitor if the building is subsequently sold. It should be noted that a local authority cannot be fined for contravening this regulation (see regulation 22 in Appendix 1).

5.4.8 Notice of energy rating

Regulation 16 requires that where a new dwelling is created, either by new building work or by a material change of use, the person carrying out the building work must calculate the energy rating of the dwelling by the Government approved Standard Assessment Procedure (SAP) and must give the local authority notice of this within 5 days of completion of the dwelling or at least 5 days before occupation if this occurs before completion. A notice stating the energy rating for the dwelling must be affixed in a conspicuous place in the dwelling by whoever is carrying out the building work, unless they intend to occupy the dwelling themselves. A similar notice procedure exists if an approved inspector is used under *regulation 12 of the Building (Approved Inspectors etc) Regulations 2000* (see Section 5.6.6).

5.4.9 Testing of building work and sampling of materials

Regulations 18 and 19 of the *Building Regulations* 2000 (as amended) empower local authorities to test building work to ensure compliance with the requirements of regulation 7 (*Materials and workmanship*) and any applicable parts of Schedule 1, and to take samples of materials to be used in the carrying out of the building work. The power to test all building work covered by the regulations was an extension of the former powers which applied only to the testing of drains and sewers. This necessitated amending regulation 18 (*see the Building (Amendment) Regulations 2001 SI 2001/3335*), and was needed so that local authorities could enforce the airtightness and insulation continuity requirements of Part L (*Conservation of fuel and power*) of Schedule 1 which were brought into effect in April 2002. Since then further requirements for carrying out sound insulation testing have been introduced by virtue of regulation 20A, *Sound insulation testing* in the *Building (Amendment)(No 2) Regulations* 2002. It should be noted that regulations 18 and 19 do not apply where the work is supervised by an approved inspector; however, similar powers exist in the *Building (Approved Inspectors etc) Regulations 2000* (as amended) (see Section 5.8.2).

Under regulation 18 local authorities '*may make such tests of any building work to establish whether it complies with regulation 7 or any applicable requirements contained in Schedule 1.*'

The regulation refers to '*tests of any building work*' but in practice testing of actual building work to establish satisfactory compliance under regulation 18 would normally be restricted to the following:

- airtightness/watertightness of drains and sewers in paragraph H1 of Part H
- airtightness and insulation continuity of large non-domestic buildings in paragraph L2 of Part L.

Interestingly, for sound insulation testing under regulation 20A, the onus is placed on the person carrying out the work to ensure that tests are carried out and the results passed to the Local Authority (or Approved Inspector).

Of course, this does not prevent a local authority from asking for test evidence that certain building materials, components or designs, etc. will achieve the necessary standards of performance to comply with the regulations. For example, manufacturer's test data could be supplied by a designer to prove that the designer met the requirements for fire resistance and surface spread of flame for a material or component, and it is common to supply test data for strength of materials, such as cube tests for concrete and pull-out tests for fixings to cladding panels.

This latest version of regulation 18 is based on (and uses very similar language to) regulation 18 from the 2000 Regulations which only applied to the testing of drains and private sewers. Under this regulation it was common for local authorities to carry out drain tests themselves with very low cost implications which could be borne by the normal local authority charges for the building control service.

It is important to realise that the onus in both the old and new versions of regulation 18 is on the local authority to make the test. For air-leakage tests which could cost between £1000 and £5000, this raises a number of questions:

(a) *Who is responsible for carrying out the test?*
It would appear that the local authority is responsible since regulation 18 mentions no other parties.

(b) *Who is responsible for paying for the test?*
Again, logic would dictate that since no other parties are referred to in regulation 18 the local authority should bear the cost if it wants a test to be carried out. Interestingly, this has often been the case with drains testing where it is not unusual for the local authority to carry out the test itself, although it is equally common for the contractor to set up the test for the local authority to inspect.

(c) *What happens if a contractor refuses to carry out the test and insists that it is the local authority that may make the test?*
There would appear to be no compulsion on a contractor to carry out a test if it did not want to. Refusal might be seen by the local authority to imply possible non-compliance and it may then wish to carry out an air-leakage test.

(d) *What happens if the local authority pays for the test to be carried out and the building subsequently passes?*

Since there would be no evidence of contravention, the local authority would have no power to take action against the contractor and consequently would be unable to recover the cost of the test.

(e) *What happens if the local authority pays for the test to be carried out and the building subsequently fails?*
In this case the contractor would be in contravention of the regulations and could be compelled by the local authority (section 36 *Building Act 1984*) to carry out the works to ensure compliance. If the contractor did not object to carrying out remedial works, it is difficult to see how the local authority could recover the costs of the original test. However, it is likely that it would require evidence that the remedial works had been effective and a subsequent air-leakage test could possibly be regarded as part of the remedial works.

Therefore, under the regulations as they are currently worded, there would seem to be nothing to prevent contractors from to refusing to carry out air-leakage tests where they were convinced that they had taken sufficient measures to comply with the functional requirements of paragraph L2 of Part L and were confident that the building would pass any air-leakage test imposed by the local authority. Even in the event of failure it is unlikely that they could be compelled to pay for the original test.

5.4.10 Sound insulation testing

The difficulties described above concerning air-leakage testing will not arise in the case of sound insulation testing, since by virtue of regulation 20A introduced by the *Building (Amendment) (No 2) Regulations 2002*, an obligation is placed on the person carrying out the work to ensure that (in the case of dwelling houses, flats and rooms used for residential purposes) appropriate sound insulation testing has been carried out to ensure compliance with Regulation E1 (*Protection against sound from other parts of the building and adjoining buildings*). The results of the test must be carried out and recorded in a manner approved by the Secretary of State (see paragraph 1.41 of Approved Document E) and must be given to the local authority not later than 5 days after completion of the works.

The requirement for pre-completion sound insulation testing has been the subject of much debate, and the response to the consultation on Part E in 2002 showed that some sections of the building industry would prefer to develop robust standard details making testing of sound insulation for a sample of new homes on each site unnecessary. Therefore, the requirement for the local authority to be given details of the pre-completion sound insulation testing was introduced in stages to give the House Builders Federation time to gear up for the changes. Working with Office of the Deputy Prime Minister (ODPM) officials, the Federation developed robust standard details and these were the subjects of a consultation exercise by the ODPM in the latter part of 2003. Therefore, the situation is that with the exception of the requirements for pre-completion sound insulation testing of new houses and buildings containing flats, the requirements of the new Part E came into force on 1 July 2003. For new houses and buildings containing flats pre-completion testing may be done and submitted as evidence of compliance; however, an alternative has existed since 1 July 2004. From that date the requirement for testing does not apply where the person carrying out the building work notifies the local authority in accordance with the Regulations that the person is using one or more of the design details approved by Robust Details Ltd.

5.4.11 Registers of notices and certificates

Under section 56 of the *Building Act 1984* (Recording and furnishing of information), local authorities must keep a register, which is available for inspection by the public at all reasonable times, giving information about notices, certificates, and insurance cover received in connection with their duties under the approved inspector system of control described in Sections 5.5 to 5.8. By notices, it means initial notices, amendment notices, notices given where there is to be a change in the person who is carrying out the work, and public body's notices. The certificates referred to are plans certificates, final certificates, public body's plans certificates, public body's final certificates and certificates given by approved persons. All these terms are discussed in the text in Sections 5.5 to 5.9. The information must be entered into the register as soon as practicable (and not more than 14 days after the event to which it relates). Information about public body's notices and initial notices need only be kept on the register for as long as they are in force.
Regulation 30 sets out the information which must be registered as:

- the description and location of the work to which the certificate or notice relates
- the name and address of the person who signed the certificate or notice
- the name and address of the insurer who signed the declaration which accompanied the certificate or notice
- the date on which the certificate or notice was accepted, or was presumed to have been accepted.

The register must contain an index which allows any person to trace an entry by reference to the address of the land to which the certificate or notice relates.

USING AN APPROVED INSPECTOR

5.5 Background

Section 49 of the Building Act 1984 defines an 'Approved Inspector' as being a person approved by the Secretary of State or a body designated by him for that purpose. Part II of the Building (Approved Inspectors etc) Regulations 2000 (as amended) sets out the detailed arrangements and procedures for the grant and withdrawal of approval.

There are two types of approved inspector:

- Corporate bodies, such as the NHBC or HCD Building Control Ltd
- Private individuals, not firms (referred to as Non-corporate Approved Inspectors).

Approval may limit the description of work in relation to which the person or company concerned is an Approved Inspector.

5.5.1 Obtaining approval as an Approved Inspector

A private individual or corporate body wishing to carry out building control functions as an Approved Inspector must be registered with the Construction Industry Council (CIC) under rules laid down by the Secretary of State. In the first years of the approved inspector system the Secretary of State approved all corporate bodies that wished to become approved inspectors. On 8 July 1996 the CIC was designated as the body responsible for approving non-corporate approved inspectors although initially the Secretary of State reserved the right to continue to approve corporate bodies. From 1 March 1999 the CIC became responsible also for the approval of corporate approved inspectors.

The CIC must also maintain a list of inspectors that it has approved. There is no express provision for these lists to be open to public inspection, although the CIC is bound to inform the appropriate local authority if it withdraws its approval from any inspector.

In approving any inspector, either the Secretary of State or the designated body may limit the description of work in relation to which the person or body is approved. Any limitations will be noted in the official lists, as will any date of expiry of approval.

Provision is also made for the Secretary of State to keep lists of designated bodies and inspectors approved by him, and for their supply to local authorities. The secretary of State must also keep the lists up to date (if there are withdrawals or additions to the list) and must notify local authorities of these changes.

In order to be registered as an Approved Inspector a number of criteria must be met. These include the holding of suitable professional qualifications, demonstration of adequate practical experience and the carrying of suitable indemnity insurance. In accordance with the responsibilities entailed by CIC's appointment as the body designated to register Approved Inspectors, it established the CIC Approved Inspectors Register (CICAIR) to maintain and operate the Approved Inspector Register. CICAIR provides applicants with a route to qualification as an Approved Inspector and, upon qualification, full Registration facilities.

The CICAIR route to qualification for Approved Inspectors involves the following four stages of assessment:

(1) *Application*

Completion of an application form and a detailed knowledge base. The knowledge base addresses the following six key areas:

  (i) Building Regulations and Statutory Control
  (ii) Law
  (iii) Construction Technology and Materials
  (iv) Fire Studies
  (v) Foundation and Structural Engineering
  (vi) Building Service and Environmental Engineering.

(2) *Pre-qualification verification*

On receipt of an application, the CICAIR Registrar will check for gaps in experience or qualification which may disqualify the Application or cause delays at further stages.

(3) *Admissions panel*

On successful completion of pre-qualification verification, the Applicant becomes a *Candidate* for Approved Inspector. The papers are then considered by professional assessors who decide whether the Candidate has demonstrated the necessary experience and knowledge to merit a *Professional Interview*. Assessors include experts nominated from across the range of disciplines by CIC Members together with qualified Approved Inspectors.

(4) *Professional interview*

Candidates granted a Professional Interview will be seen by an Interview Panel consisting of three assessors assisted by the CICAIR Registrar. The Professional Interview is the final stage of assessment and is an opportunity for candidates to demonstrate their knowledge and experience, and expand upon the information provided in their Application.

Successful completion of the above assessment stages will result in the Candidate being invited to register as an Approved Inspector. The approval will be for a period of 5 years. Further terms of approval may be sought. CICAIR is required to maintain a list of the inspectors which it has approved. This may be obtained by accessing the CIC web site at www.cic.org.uk.

5.5.2 Termination of approval

The approval of an inspector may be withdrawn by the approving body (the CIC) by notice, for example, in writing to the inspector if the inspector has contravened any relevant rules of conduct or has shown that he or she is unfitted for the work.

More seriously, where an approved inspector is convicted of an offence under section 57 of the *Building Act 1984* (which deals with false or misleading notices and certificates, etc.) the CIC may withdraw its approval. In this case, the convicted person's name would be removed from the list for a period of 5 years. There is no provision for appeals or reinstatement.

Additionally, the Secretary of State has powers to withdraw the designation of the approving body. If this were to happen it would not affect the status of any approvals given by the designated body, although the Secretary of State retains the right to withdraw an approval if the Secretary of State deems it necessary.

5.5.3 Insurance requirements

All approved inspectors are required to carry insurance cover in accordance with a scheme approved by the Secretary of State. The insurance requirements are complex and different cover is required for different types of work. For example, at present the NHBC is the only Approved Inspector with an approved insurance scheme which permits them to carry out building control on speculative dwellings. For this, the NHBC has to provide two different types of policies:

- 10-year no-fault insurance against breaches of building regulations relating to site preparation and resistance to moisture, structure, fire, drainage and combustion appliances
- insurance against the approved inspector's liability in negligence for 15 years from the issue of the final certificate for each dwelling.

In order to carry out building control on commercial buildings, self-build dwellings, social housing for rent (including staff accommodation in hotels, hostels, nurses homes, student halls of residence, etc.) the Approved Inspector must provide professional indemnity insurance renewable on an annual basis. Cover must be provided for claims against damage (including injury) resulting from the negligent performance by the Approved Inspector who issued the initial notice (see Section 5.6.5).

5.6 Procedures

Flow chart 5.2 outlines the essential stages in the control procedure that may be used by a person intending to carry out any building work to which the regulations apply, as an alternative to using the local authority system described in the previous sections in this chapter.

5.6.1 Appointing an Approved Inspector

A list of both Corporate and Non-corporate Approved Inspectors, who are capable of supervising work under the Building Regulations, may be obtained from the CIC or the Association of Consultant Approved Inspectors (ACAI) (www.acai.org.uk).

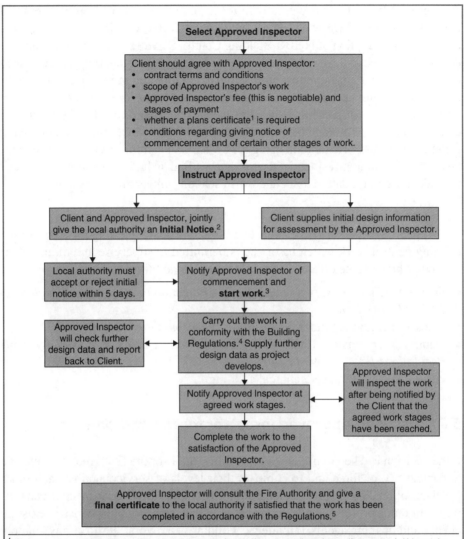

Select Approved Inspector

Client should agree with Approved Inspector:
- contract terms and conditions
- scope of Approved Inspector's work
- Approved Inspector's fee (this is negotiable) and stages of payment
- whether a plans certificate[1] is required
- conditions regarding giving notice of commencement and of certain other stages of work.

Instruct Approved Inspector

Client and Approved Inspector, jointly give the local authority an **Initial Notice**.[2]

Client supplies initial design information for assessment by the Approved Inspector.

Local authority must accept or reject initial notice within 5 days.

Notify Approved Inspector of commencement and **start work**.[3]

Approved Inspector will check further design data and report back to Client.

Carry out the work in conformity with the Building Regulations.[4] Supply further design data as project develops.

Notify Approved Inspector at agreed work stages.

Approved Inspector will inspect the work after being notified by the Client that the agreed work stages have been reached.

Complete the work to the satisfaction of the Approved Inspector.

Approved Inspector will consult the Fire Authority and give a **final certificate** to the local authority if satisfied that the work has been completed in accordance with the Regulations.[5]

[1] The person carrying out the building work may also request that the Approved Inspector supplies a plans certificate a copy of which must also be supplied to the local authority. Possession of a plans certificate can give valuable protection in the event that the initial notice is cancelled or ceases to be in force and no new initial notice is given or accepted. A plans certificate may be combined with the initial notice or given subsequently. If the Approved Inspector is unable to give a plans certificate an application may be made to the Secretary of State for a determination and a fee is payable. It may be necessary for the Approved Inspector to consult the Fire Authority before giving a plans certificate.

[2] An initial notice must be accompanied by certain details and by evidence that the Approved Inspector is suitably insured. It may be necessary for the Approved Inspector to consult the Fire Authority and the sewerage undertaker at this stage. If the initial notice is cancelled the work must not be continued with unless it is being supervised either by a new Approved Inspector or by the local authority. Another Approved Inspector may be engaged provided that the local authority has not taken positive steps to supervise the work.

[3] Work may be started once the initial notice has been expressly accepted by the local authority (or is deemed to have been accepted after 5 days have elapsed without its being rejected). Work should not be started if the notice is rejected.

[4] The work must comply with the Regulations. The Approved Inspector is empowered to give a written notice to the person carrying out the building work if he considers that the work contravenes the Regulations. Failure to remedy the contravention within 3 months of such a notice will lead to cancellation of the initial notice by the Approved Inspector.

[5] The local authority has 10 days in which to reject the final certificate on grounds specified within the Approved Inspector regulations (see Section 5.8.4).

Flow chart 5.2 The Approved Inspector system of building control.

Approved Inspectors vary greatly in terms of size (number of employees) and geographical spread (regional or national). Some offer additional services, however, with the exception of NHBC Building Control Services Ltd, which remains the only body insured to deal with speculative domestic construction (i.e. self-contained houses, flats and maisonettes built for sale to private individuals), no other approved inspectors are permitted to offer a building control service for the erection of such dwellings at present. However, they are permitted to offer their building control services for a range of other types of 'dwellings' including social housing for renting tenants, self-build housing for owner occupation, service flats functionally connected to non-residential uses (e.g. staff flats in hotels, pubs, etc.) and staff or student accommodation associated with schools, universities, hospitals, etc.

Scope of work

The Approved Inspector will be able to advise the client regarding the scope of the building control work which he or she is permitted to supervise and of any other services which can be offered. The most common additional services involve:

- fire safety risk assessments (under the *Fire Precautions (Workplace) Regulations 1997 (SI 1997/1840)*)
- accessibility audits (under the *Disability Discrimination Act 1995*)
- planning supervision (under the *Construction (Design and Management) Regulations 1994 (SI 1994/3140)*)
- latent defects insurance technical auditing.

5.6.2 Using an Approved Inspector for the first time

If the decision is taken to use an Approved Inspector for the first time, it is sensible to contact two or three and to compare their levels of service against each other and those offered by the relevant local authority. Some Approved Inspectors can provide continuing professional development (CPD) sessions for prospective or existing clients and/or their agents (architects, building surveyors, etc.) on any matters related to building control legislation and some will provide this free of charge as a way of getting to know new clients. Of course, they will want to make use of such an opportunity to present their services to the client and his or her design team.

The client can request that each selected Approved Inspector provide him with a 'fee bid' for a specific contract. Approved Inspector's fees are negotiable since there are no statutory fees laid down by the Government. For individual contracts, most Approved Inspectors base their fees on the cost of the regulated work in the project. Therefore, in order to obtain comparable fee bids for the same project, the same contract sum should be quoted to each competing bidder (Approved Inspectors and Local Authority).

As with most areas of business activity, the laws of supply and demand apply to the extent that a client who can show that a large continuing portfolio of work will be available to the successful bidder may expect to negotiate fees on more advantageous terms than if the work is presented on a piecemeal basis. If such a client operates on a wide regional, or even national, basis then it may be more advantageous to approach an Approved Inspector who is similarly based. In these circumstances, the Approved Inspector would expect to enter into a 'term contract' with the client whereby all future contracts requiring Building Regulation assessment work would be allocated to the Approved Inspector for a given period of time. However, experience has shown that most clients prefer to judge the Approved Inspector's performance on a single specific project before entering into a lengthier arrangement. In addition to reduction of fees, the main advantage of a term contract is that it speeds up the process of appointment and avoids the need for having to seek fee bids and agree contract terms on each and every contract.

5.6.3 Instructing an Approved Inspector

When instructing an Approved Inspector the client should be aware that a formal contractual arrangement will exist, and all arrangements between the parties will be subject to the law of England and Wales. Essentially, the Approved Inspector is acting as a consultant who is being commissioned by the client to carry out a Building Regulations assessment service. Therefore, it is usual for the contract to contain clauses covering:

- The consultant's services and obligations, that is,
    - (i) the exercise of reasonable skill, care and diligence
    - (ii) the use of appropriately qualified and experienced staff
    - (iii) the provision of appropriate tools, materials and equipment
    - (iv) the use of subcontractors
    - (v) the procedures to be adopted regarding Initial Notices, Plans Certificates and Final Certificates
    - (vi) the need for consultations with relevant bodies (e.g. the fire authority)
    - (vii) liability for loss or damage arising directly or indirectly from the consultant's performance
    - (viii) the need to keep proper records of activities and transactions for which payment is based on time charges.
- The client's requirements and obligations, that is,
    - (i) the form of the client's instruction to the consultant
    - (ii) the supply of information
    - (iii) the need to inform the consultant at commencement and at any other agreed stages of work
    - (iv) site access arrangements and facilities for testing services and sampling materials
    - (v) remedies for obstruction or delay
    - (vi) occupation of the completed works
    - (vii) payment conditions.

- Intellectual property rights and confidentiality of both parties to the contract.
- Details of the consultant's insurance requirements.
- The status of the consultant regarding conflicts of interest. Except for minor work (see Section 5.6.4), the Approved Inspector is not permitted to have a professional or financial interest in the building work.
- Questions arising between the consultant and the client regarding conformity of the plans with the Building Regulations. Normally, any disputes would be referred to the Secretary of State for a determination (see Chapter 6).
- The methods for dealing with contraventions, relaxations and dispensations (see Chapter 6).
- Methods for dealing with the transmission of notices under the contract.

Level of service

The level of service offered may vary between Approved Inspectors, but it will always meet the standards laid down in the Building Control Performance Standards (see Section 5.4.1) for Approved Inspectors who are members of the ACAI. Therefore, all such Approved Inspectors will check that the design complies with:

- all relevant parts of the Building Regulations (including carrying out checks on structural, thermal and fire engineering design calculations, if these form part of the design details)
- all sections of Acts of Parliament which constitute powers linked to Building Regulations under which plans can be rejected.

Additionally, there are certain statutory provisions (London Building Acts, Local Acts of Parliament, licensing provisions, Safety of Sports Grounds legislation, etc.), which remain the responsibility of the relevant local authority to enforce. Requirements laid down by local authorities under such legislation can affect Building Regulations assessment decisions made by the Approved Inspector. Therefore, most Approved Inspectors will offer to negotiate with the relevant local authority on behalf of the client for compliance with such provisions. As this service will normally attract an additional fee, it should be referred to explicitly in the contractual agreement with the Approved Inspector.

Usually, the Building Regulations assessment service can be tailored to meet the client's (or the clients' agent's) (a) requirements, (b) programme of work and (c) the methods of working. This can include:

- the provision of assessments and reports at all design stages from sketch scheme to the production of site working drawings
- being able to receive information in stages, as and when it is produced by the designers
- the rapid turn-round of assessment reports when speedy design decisions have to be made
- the assessment of design revisions irrespective of when they are produced
- providing type approvals for repetitive designs

- the assessment of innovative designs and non-standard design solutions
- the ability to handle drawings in electronic as well as hard-copy forms
- carrying out site inspections of work in progress to suit the contractors schedule of work
- arranging for the payment of fees to suit the clients financial planning.

This ensures that building control is seen as an integral part of the design process rather than as a hurdle that has to be overcome at inconvenient intervals.

Formal instruction

Once negotiations have been completed and the contract arrangements have been settled the Approved Inspector will require a formal instruction from the client. Most Approved Inspectors can provide the client with a standard Instruction Notice which when received and acknowledged by the Approved Inspector formally establishes the contractual relationship between the parties for each particular project (or series of projects) as appropriate.

5.6.4 Minor work

An Approved Inspector must have no professional or financial interest in the work supervised, unless it is minor work. For example, this prevents an Approved Inspector from supervising work that he or she has designed or constructed, or from working for a company which has an interest in the work.

Minor work includes:

- The extension or alteration of a single- or two-storey dwelling-house (or its controlled services or fittings), provided that the dwelling-house does not exceed three storeys after the alterations (any basements may be disregarded when counting storeys). The definition of dwelling-house does not include flats.
- Work involving the underpinning of a building.

5.6.5 The Initial Notice

When the client has appointed an Approved Inspector for a particular project, the first step in the Approved Inspector process will be the service of an **Initial Notice** on the local authority in whose area the work is to be carried out. This will make the local authority aware that building work in their area is being legally controlled under the Regulations, and will notify them of certain linked powers that they have under the *Building Act 1984* and any local Acts of Parliament. The local authority has 5 days in which to accept or reject the initial notice. Acceptance effectively suspends the local authority's powers to enforce the Building Regulations for the work described in the notice for as long as it remains in force, and the supervisory function passes to the Approved Inspector.

Thus the initial notice is crucial to the operation of the Approved Inspector system, and great care must be taken that:

- it is completed correctly
- it is served on the correct office of the relevant local authority
- it is accepted by the local authority before any work starts on site.

The notice must be in a form prescribed by the *Building (Approved Inspectors etc) Regulations 2000 (SI 2000/2532), regulation 12, Form 1* as amended by the *Building (Approved Inspectors etc)(Amendment) Regulations 2000 (SI 2001/3336)*. To simplify completion of the initial notice, much of the information required by law can be pre-printed. Indeed, it is a usual practice for the Approved Inspector to supply the client with a quantity of blank notices which can then be filled out with the information specific to the client and the proposed building work as the need arises. The partially completed notice is then returned to the Approved Inspector for completion and forwarding to the Local Authority.

A typical initial notice is illustrated in Figure 5.3. The parts, which should be completed by the client, are as follows:

(a) Box 2 (*This notice relates to the following work*):
  – Give the address of the site
  – Give a brief description of the work (e.g. the erection of a single-storey extension; the erection of a four-storey detached building)
  – Give details of the use of the building and the use of any building affected by the work (e.g. office, factory, hotel, etc.).
(b) Box 3 (*The person intending to carry out the work is*):
  – Give the name and address of the client (not the architect, contractor, etc.).
(c) Box 6 (*Allied Approved Inspectors Ltd declares that*):
  – The client (or his legal representative) should sign at the bottom of this box
  – Give the name and position in the client's organisation of the person signing the notice
  – Date the form.

Additionally, the client (or Client's agent) will need to furnish the Approved Inspector with certain information pertaining to the project which will need to be returned with the notice. This information is referred to in Box 5 and includes the following:

(a) Where the work relates to the erection or extension of a building, a site plan to a scale of not less than 1:1250 which shows:
  – the boundaries and location of the site.
(b) Where the work includes the construction of a new drain or private sewer, an indication within the site plan and/or an accompanying statement:
  – as to the approximate location of any proposed sewer connection (for both foul and surface water drainage) or

Allied Approved Inspectors Ltd
Initial notice

This notice is issued pursuant to section 47 of the Building Act 1984 ('the Act') and the Building (Approved Inspectors etc) Regulations 2000 ('the 2000 Regulations') as amended

FORM 1

Box 1
To: (Note 1)
....
....
....
....

Box 2
This notice relates to the following work: (Note 2)
Location: ....
....
Description of work: ....
....
Use of any building: ....

Box 3
The person intending to carry out the work is: (Note 3)
Name: ....
Address: ....
....
....Postcode.........................

Box 4
The Approved Inspector in relation to the work is: Allied Approved Inspectors Ltd, 1 Any Street, Anytown, AB12 3CD
A declaration signed by the insurer for Allied Approved Inspectors Ltd accompanies this notice and states that a named scheme of insurance approved by the Secretary of State applies in relation to the work described. Allied Approved Inspectors Ltd was approved under section 3(1) of the 2000 regulations and a copy of the notice of approval accompanies this notice.

Box 5
With this notice are the following documents, which are those relevant to the work described in this notice: (Note 4)
(a) in the case of the erection or extension of a building, a plan to a scale of not less than 1:1250 showing the boundaries and location of the site, and where the work includes the construction of a new drain or private sewer, an indication within that plan and/or an accompanying statement:
(i) as to the approximate location of any proposed connection to be made to a sewer, or
(ii) if no connection is to be made to a sewer, as to the proposals for the discharge of the proposed drain or private sewer including the location of any septic tank and associated secondary treatment system, or any wastewater treatment system or any cesspool. (Note 5)
(b) the following **Local Enactment** is relevant to the work (*delete if none applies*): -
....
The following steps will be taken to comply with it:-
....
....

Figure 5.3 Typical initial notice.

Allied Approved Inspectors Ltd
Initial notice

Allied Approved Inspectors Ltd declares that: | Box 6 |

(i)   The work is/is not* minor work (Note 6)

(ii) It does not and will not while this notice is in force have any financial or professional interest in the work described. (Note 7)

(iii) It will/will not* be obliged to consult the fire authority by regulation 13 of the 2000 Regulations. (Note 8)

(iv) It undertakes that before giving a plans certificate in accordance with section 50 of the Act, or a final certificate in accordance with section 51 of the Act, it will consult the fire authority in respect of any work described above. (Note 9)

(v)   It will/will not* be obliged, by regulation 13A of the AI regulations, to consult the sewerage undertaker. (*If the sewerage undertaker has to be consulted the next following declaration must be made*)

(vi) It undertakes that before giving a plans certificate in accordance with section 50 of the Act, or a final certificate in accordance with section 51 of the Act, it will consult the sewerage undertaker in respect of any of the work described above. (Note 9)

(vii) It is aware of the obligations laid upon it by Part II of the Act and by regulation 11 of the 2000 Regulations.

(*delete whichever statement is inapplicable)

Signed..   Name...

For and on behalf of Allied Approved Inspectors Ltd   Position...
Approved Inspector

Date...

Signed..   Name...

For and on behalf of the person intending to carry   Position...
out the work (Note 10)

Date...

Notes

(1) Insert the name and address of the local authority in whose area the work will be carried out. This should be addressed to the office of the building control department, not the general local authority address. Failure to insert the correct address may lead to delays in acceptance of this notice.
(2) Location and description of the work and the use of any building to which the work relates.
(3) Insert the name and address of the person intending to carry out the work (this will usually be the client who commissions the work, not his agents or contractors, etc.).
(4) The local authority may reject this notice only on grounds prescribed by the Secretary of State. These are set out in Schedule 3 to the 2000 Regulations. They include failure to provide the relevant documents set out in this section. The documents listed in this section of the notice relevant to the work described above should therefore be sent with this notice. Any sub-paragraph which does not apply should be deleted.
(5) The design of any drainage system to which the requirements of Part H of Schedule 1 to the Building Regulations 2000 apply, and which therefore falls to be considered by the Allied Approved Inspectors Ltd, will not necessarily be shown in full on the plans accompanying this notice. The plans will indicate the *location* of any connection to be made to a sewer, or the proposals for the *discharge* of any proposed drain or private sewer including the *location* of any septic tank and associated secondary treatment system, or any wastewater system or any cesspool.

Figure 5.3 (Continued).

> (6) 'Minor work' has the meaning given in regulation 10(1) of the 2000 Regulations. If the work is **not** minor work, the next following declaration **must** be made.
> (7) 'Professional or financial interest' has the meaning given in regulation 10 of the 2000 Regulations.
> (8) If the inspector is obliged to consult the fire authority, the next following declaration **must** be made.
> (9) Delete this statement if it does not apply.
> (10) The person intending to carry out the work will usually be the client who commissions the work, not his agents or contractors, etc.

Figure 5.3 (Continued).

- proposals for disposing with foul and surface water drainage if no sewer connection is to be made (e.g. the location of any septic tank, and associated secondary treatment system, or any wastewater treatment system or cesspool).

Although the initial notice specifies that the information under (b) above can be in the form of a statement, it is usual (as well as being clearer and more convenient) to provide this information on the 1:1250 scale site plan referred to under (a). Full details of any drainage design to which the requirements the *Building Regulations 2000 (SI 2000/2531), Schedule 1, Part H* apply do not have to be provided to accompany the initial notice, as this is part of the Building Regulations assessment which the Approved Inspector has been commissioned to carry out.

Once the Approved Inspector has received the notice and the information referred to above from the client, the approved inspector will be able to complete the notice and send the whole package to the relevant local authority. The Local Authority must accept or reject the notice within 5 days of receipt (if they do not do this the notice becomes accepted by default). Therefore, the Approved Inspector will send the notice and accompanying documents by recorded delivery so that proof of receipt can be verified.

An Approved Inspector may feel that if a rejection has not been received within the 5-day period then an assumption of acceptance can be made. In practice, problems have been encountered whereby the initial notice has been sent to the principal address of the local authority, and has then taken some time to be forwarded to the correct area building control office. In these circumstances the area office may assume that the 5-day period starts on the day the notice arrived at their particular office. To avoid these problems, the *Manual to the Building Regulations* (published by the ODPM) recommends that '*it is important that the initial notice is addressed to the correct part of the local authority and the correct building*'.

Change of client during the construction period

It may happen that the work to which an initial notice relates is to be carried out by a different person to the one who gave the initial notice with the Approved Inspector. This can be achieved if the Approved Inspector and a person who proposes to carry out work in succession to the person who gave the initial notice, jointly give a written notice to that effect to the Local Authority. The initial notice is then treated as having been given by the Approved Inspector and the new person intending to carry out the work.

Cancellation of the initial notice

Generally, the initial notice remains in force during the currency of the works. However, in certain circumstances, it may be cancelled, or cease to have effect after the lapse of certain defined periods of time where there has been a failure to give a Final Certificate (see Section 5.8.4) to the local authority.

In the following cases, the Approved Inspector must cancel the initial notice by issuing to the Local Authority a cancellation notice in a prescribed form:

- The Approved Inspector has become or expects to become unable to carry out (or continue to carry out) his or her functions.
- The Approved Inspector believes that because of the way in which the work is being carried out he or she cannot adequately perform his or her functions.
- The Approved Inspector is of the opinion that the requirements of the Regulations are being contravened and despite giving notice of contravention to the person carrying out the work that person has not complied with the notice within the 3-month period allowed.

It is also possible for the person carrying out the work to cancel the initial notice. This arises if it becomes apparent that the Approved Inspector is no longer willing or able to carry out his or her functions (through bankruptcy, death, illness, etc.). This must be done in the prescribed form and must be served on the local authority and (where practicable), on the Approved Inspector. A typical form is shown in Figure 5.4.

Alternatively, it is possible for the person carrying out the work to give a new initial notice jointly with a new Approved Inspector, provided that the new notice is accompanied by an undertaking by the original Approved Inspector that he or she will cancel the earlier notice as soon as the new notice is accepted.

The initial notice will automatically cease to have effect after the lapse of certain defined periods of time, where there is a failure to give a final certificate (see Section 5.8.4).

Once the initial notice has ceased to have effect, the Approved Inspector will be unable to give a final certificate and the Local Authority's powers to enforce the Building Regulations can revive. This has the consequences described in Section 5.8.3.

Finally, a local authority may cancel an initial notice if it appears to them that the work to which the initial notice relates has not been commenced within a period of 3 years dating from when it was accepted by the local authority. There is a prescribed form for this (see Form 8, *Notice of cancellation by local authority*) in Schedule 2 of the *Building (Approved Inspectors etc) Regulations 2000*. The notice must be given to the approved inspector who gave the initial notice and to the person intending to carry out the work as shown on the initial notice.

FORM 7

Section 52(3) of the Building Act 1984 ('the Act')

The Building (Approved Inspectors etc) Regulations 2000 ('the 2000 Regulations')

NOTICE OF CANCELLATION BY PERSON CARRYING OUT WORK

To: (Note 1)

....
....
....

1.   This notice relates to the following work: (Note 2)

Location:...
....
....

Description of work:..
....
....
....

Use of any building:..
....

2.   An initial notice dated...............(Note 3) has been given and the above work was specified in it.

3.   I am the person carrying out the work/intending to carry out the work. (Note 4)

4.   I hereby cancel the initial notice.

Signature..

Date.........................

Notes
(1) Insert the name and address of the person to whom the notice is given. It must be given to the local authority and, if practicable, to the Approved Inspector.
(2) Location and description of the work including the use of any building to which the work relates.
(3) Insert date.
(4) Delete whichever does not apply.

Figure 5.4 Typical notice of cancellation by the person carrying out the work.

Local authority's powers to reject the initial notice

The local authority can reject the notice only if it is defective in terms of accuracy and completeness. The grounds for rejection are:

• the notice is not in the prescribed form
• the notice has been served on the wrong local authority

- the person who signed the notice as an Approved Inspector is not an Approved Inspector
- the information supplied is deficient because neither the notice nor plans show the location or contain a description of the work (including the use of any building to which the work relates)
- the Approved Inspector is obliged to consult the sewerage undertaker before giving a plans certificate or final certificate, and the initial notice does not contain an undertaking to do so
- where it is intended to erect or extend a building, the local authority considers that a proposed drain must connect to an existing sewer, but no such arrangement is indicated in the initial notice
- the initial notice is not accompanied by a copy of the Approved Inspector's notice of approval
- evidence of approved insurance is not supplied
- the notice does not contain an undertaking to consult the fire authority (where this is appropriate)
- the notice does not contain a declaration that the Approved Inspector has no financial or professional interest in the work (this does not apply to minor work)
- local legislative requirements will not be complied with
- an earlier initial notice has been given for the work, which is still effective. This ground for rejection does not apply if:
  (i) the earlier notice has ceased to be in force and the Local Authority has taken no positive steps to supervise the work described in it
  (ii) the initial notice is accompanied by an undertaking from the Approved Inspector who gave the earlier notice such that the earlier notice will be cancelled when the new notice is accepted.

It is most important that the local authority accepts the initial notice before the work commences on site since any substantive work carried out before acceptance could be regarded as unregulated and in contravention of the Building Regulations. Therefore, sufficient time should be allowed in the client's programme for the processes detailed above. On rare occasions there have been instances where work has started before or during the 5-day period and the local authority has rejected the initial notice on this basis. Clearly, this course of action would appear to be illegal since premature commencement of the work is not included in the grounds for rejection. If building work has been carried out before acceptance of the initial notice and becomes the subject of local authority enforcement action, any other work on the same project covered by the initial notice which had not been commenced, could still be supervised by an Approved Inspector.

5.6.6 Ensuring compliance with the Building Regulations

When the initial notice has been accepted, the Approved Inspector is required by regulation 11 of the *Building (Approved Inspectors etc) Regulations 2000* (as amended) to '*take such steps as are reasonable to enable him to be satisfied within the limits of professional skill and care that*':

(i) Building work complies with the Building Regulations.
(ii) The regulations concerning the calculation of energy ratings for dwellings, are complied with.
(iii) The regulation concerning sound insulation testing, is complied with (this requirement comes into force in stages as described in Section 5.4.9).
(iv) Where building work involves the insertion of insulating material into the cavity in a wall after that wall has been built, the Approved Inspector is not required to supervise the insertion of the material, but must state in the final certificate (see Section 5.8.4) whether or not the material has been inserted.

In theory, each Approved Inspector can make whatever procedural arrangements he or she considers necessary to meet these requirements. In practice, the *Building Control Performance Standards* make recommendations covering the provision of a plan appraisal service and require the adoption of an appropriate site inspection regime, and the Regulations require, where appropriate, that the Approved Inspector consults the fire authority and the sewerage undertaker at certain stages during the execution of the works.

Certain obligations regarding the supply of information are also placed on the person carrying out the building work. For example, where a new dwelling is created by building work or by a material change of use, the client (i.e. the person carrying out the building work) is under an obligation to calculate the energy-rating of the dwelling and give notice of it to the Approved Inspector. This must be done not later than 5 days after completion of the dwelling where it is created by building work or, if it is occupied before completion, not later than 8 weeks from the date of occupation.

Where the dwelling is created by a material change of use, the notice must be given not later than 8 weeks from the date on which the change of use took place.

Additionally, the person carrying out the building work is obliged either to display the notice of energy rating in a conspicuous place in the dwelling or to give a copy of it to the occupier, although this does not apply where the person carrying out the building work is also the occupier (i.e. for self-builders). Failure to provide the Approved Inspector with the energy rating notice, or to display it in the dwelling (or give it to the occupier, as appropriate), could result in the Approved Inspector being unable to give the final certificate. This could result in the consequences described in Section 5.8.4.

The contract between the client and the Approved Inspector will lay down the ground rules for design assessments, site inspections, fire authority consultations and consultations with the sewerage undertaker. The working arrangements, however, will be agreed on a job-by-job basis with the client's design team and contractor in order to fit in with the design and construction programmes.

In order to ensure that design assessments, site inspections, fire authority consultations and consultations with the sewerage undertaker are properly recorded the Approved Inspector will always follow up each event with a written report to the relevant responsible person in the client's organisation. These reports will indicate areas of non-compliance, pass on the views of consultees (fire authority, statutory undertakers, etc.) and explain the remedies available should there be a dispute over compliance.

Design assessment and approval of plans

The design assessment service provided by the Approved Inspector is covered by the *Building Control Performance Standards* which recommend that clear information should be communicated by the Approved Inspector to the client regarding:

- Non-compliance with the Building Regulations.
- Views of statutory undertakers.
- Any conditions (such as the provision of information by specific dates) pertaining to the informal approval of plans.
- Remedies available in the event of a dispute over compliance (such as applying to the ODPM for a determination).

Additionally, it is recommended that the Approved Inspector keep records of the design assessment philosophy and of any statutory and/or discretionary consultations, to enable satisfactory continuity of control and for future reference.

Most Approved Inspectors encourage the client to provide design information on a continuous basis and they will furnish the client with regular reports on the progress of the design assessment. There are no statutory 'approval' deadlines to be met and the Approved Inspector will fit in with the normal design programme thus providing a continuous checking service.

Most clients like to be reassured that their designs do, in fact, comply with the relevant regulations. This can be done informally by the provision of a letter to that effect from the Approved Inspector, once the design has reached a stage from which it is unlikely to be substantially altered.

5.6.7 The Plans Certificate

If a formal reassurance is required, the Approved Inspector can, at the request of the client, issue a '**Plans Certificate**' to the local authority. In order to issue a plans certificate, the Approved Inspector will need to certify that:

- the client's plans have been checked
- he or she is satisfied that they comply with the Building Regulations
- any prescribed consultations (e.g. with the fire authority, sewerage undertaker, etc. if appropriate) have been carried out.

The certificate must be endorsed with the reference numbers of the inspected plans (although copies of the actual plans do not have to be sent in with the certificate). Other than the provision of the plans in the first place, the client has no input into the preparation and transmission of the plans certificate.

From the client's viewpoint, the advantage of possessing a plans certificate lies in the fact that if at a later stage the initial notice ceases to be effective, the local authority cannot take enforcement action in respect of any work described in the plans certificate if it has been done in accordance with those plans.

A plans certificate, when issued by an Approved Inspector, certifies that the design has been checked and that the plans comply with the Building Regulations. Its issue is entirely at the option of the person carrying out the work, and copies of it are sent by the Approved Inspector to that person and to the Local Authority.

If an Approved Inspector declines to issue a plans certificate on the grounds that the plans do not comply with the Regulations, the building owner can refer the question of compliance to the Secretary of State for a determination (see Chapter 6). Similarly, if the client believes that the requirements of the Building Regulations are too onerous taking into account the particular circumstances of the case, they may apply to the relevant local authority for a relaxation or dispensation of those regulations which appear too onerous (see Chapter 6). This applies irrespective of whether an Approved Inspector or Local Authority has been used in the first instance.

A plans certificate, which must be in a prescribed form, can be issued at the same time as the initial notice, or at a later stage, provided that the work referred to has not been carried out.

5.6.8 Local authority's powers to reject a plans certificate

The local authority has 5 working days in which accept or reject the plans certificate, but it may only reject it on certain specified grounds:

- The plans certificate is not in the prescribed form.
- It does not describe the work to which it relates.
- It does not specify the plans to which it relates.
- Unless it is combined with an initial notice, that no initial notice is in force.
- The plans certificate is not signed by the Approved Inspector who gave the initial notice or that he or she is no longer an Approved Inspector.
- Evidence of approved insurance is not supplied.
- The certificate does not contain a declaration that the fire authority has been consulted (where this is appropriate).
- The Approved Inspector was obliged to consult the sewerage undertaker before giving the certificate, but the certificate does not contain a declaration that he or she has done so.
- There is no declaration of independence (except for minor work).

If the plans certificate is combined with an initial notice, the grounds for rejecting an initial notice (see Section 5.6.5) also apply. Plans certificates may be rescinded by a local authority if the work has not started within 3 years of the acceptance date of the certificate. A typical Plans Certificate is shown in Figure 5.5.

FORM 3

Section 50 of the Building Act 1984 ('the Act')

The Building (Approved Inspectors etc) Regulations 2000 ('the 2000 Regulations')

PLANS CERTIFICATE

1 This certificate relates to the following work(Note 1)
   ....
   ....
   ....

2 I am an approved inspector for the purposes of Part II of the Act and the above work is [the whole]/[part](Note 2) of the work described in an initial notice given by me and dated.(Note 3)

3 With this certificate is the declaration, signed by the insurer, that a named scheme of insurance approved by the Secretary of State applies in relation to the work to which the certificate relates.

4 Plans of the work specified above have been submitted to me and I am satisfied that the plans neither are defective nor show that work carried out in accordance with them would contravene any provision of building regulations.

5 The work [is]/[is not](Note 2) minor work.(Note 4)

6 I declare that I have no financial or professional interest(Note 5) in the work described since giving the initial notice described in paragraph 2.(Note 6)

7 I have consulted the fire authority in accordance with regulation 13.(Note 6)

7A I have consulted the sewerage undertaker in accordance with regulation 13A.(Note 6)

8 The plans to which this certificate relates bear the following date and reference number: (Note 7)

....
....
....

Signed..

Approved Inspector

Date........................

Notes
(1) Location and description of the work including the use of any building to which the work relates.
(2) Delete whichever does not apply.
(3) Insert date.
(4) 'Minor work' has the meaning given in regulation 10(1) of the 2000 Regulations. If work is not minor work, the declaration in paragraph 6 must be made.
(5) 'Professional or financial interest' has the meaning given in regulation 10 of the 2000 Regulations.
(6) Delete this statement if it does not apply.
(7) Insert the date and reference number.

Figure 5.5 Typical plans certificate.

5.6.9 Dealing with a change in the design

Design changes which are unlikely to alter the description or scope of the work given in the initial notice are usually dealt with by means of revised plans submitted to the Approved Inspector at intervals during the course of the design or site works. The Approved Inspector will check the plans and issue reports to the client in accordance with the agreed contractual arrangements. For more extensive design changes that materially alter the scope of the works (e.g. it may be decided to change the number of units being erected on a site, or the scale of the work may alter substantially) it may be necessary for the Person who is carrying out the work and the Approved Inspector jointly to give an amendment notice to the Local Authority. There is a prescribed form for an amendment notice and it must contain the information which is required for an initial notice (see Section 5.6.5) plus either:

- a statement to the effect that all plans submitted with the original notice remain unchanged or
- copies of all the amended plans with a statement that any plans not included remain unchanged.

The local authority has 5 working days in which to accept or reject the notice, and it may only reject it on prescribed grounds. The procedure is identical to that for acceptance or rejection of an initial notice (see Section 5.6.5). Where copies of amended plans must be supplied, it is assumed that these are similar in content to those submitted with the original initial notice; that is, where the work relates to the erection or extension of a building, a site plan to a scale of not less than 1:1250 which shows the boundaries and location of the site. There would be little point in submitting detailed plans of the revised design since the local authority would have no powers to inspect these, and would not have been supplied with the original design details in any case. However, there is no definition of 'plans' in the Approved Inspector regulations.

5.7 Consultations

5.7.1 Fire authority consultation

Where an initial notice or an amendment notice is to be given (or has been given) in relation to the erection, extension, material alteration or change of use of a building which:

- is to be put to a designated use under the *Fire Precautions Act 1971, section 1*, or
- will be a workplace subject to Part II of the *Fire Precautions (Workplace) Regulations 1997 (SI 1997/1840)*

and the *Building Regulations 2000 (SI 2000/2531), Schedule 1, Part B (Fire safety)* also applies, the Approved Inspector is required, before or as soon as practicable after giving the notice, to consult the fire authority. The Approved Inspector must give them sufficient plans, and/or other information to show that the work described in the notice will comply with the applicable parts of the *Building Regulations 2000 (SI 2000/2531), Schedule 1, Part B* and must have regard to any views they express.

Additionally, before giving a plans certificate or final certificate to the Local Authority the Approved Inspector must allow the Fire Authority 15 working days to comment, and have regard to the views they express.

Some local Acts of Parliament also impose extensive Fire Authority consultation requirements. The Approved Inspector must undertake any consultation required by local legislation.

In the past, there has been some confusion regarding the legal status of any advice offered by the fire authority during the consultation process. In the *Building (Approved Inspectors etc) Regulations 2000 (SI 2000/2532), regulation 13* (Approved Inspector's consultation with the Fire Authority), paragraph 4(a) makes it clear that the Approved Inspector must provide '*sufficient plans to show whether the work would, if carried out in accordance with those plans, comply with the applicable requirements of Part B of Schedule 1 to the Principal Regulations*'. Therefore, the phraseology of paragraph 4(a) is a means of describing plans that adequately cover those features of the work that are of relevance to fire safety. This point is reinforced in *DETR Circular 07/00* dated 13 October 2000, which accompanied the issue of the *Building (Approved Inspectors etc) Regulations 2000 (SI 2000/2532)*. Paragraph 6.8 of the Circular states '*There is no longer an implication that the fire authority have an authoritative view on the compliance of building work with Part B, although it is of course open to them to offer informal views on that matter. The primary object of the consultation is to provide an opportunity for the approved inspector and the fire authority to reach mutually compatible views on whether plans of the building work are satisfactory from the standpoints of the building regulations and of fire precautions legislation.*'

5.7.2 Consultations with the sewerage undertaker

Where an initial notice or amendment notice is to be given (or has been given) and it is intended to erect, extend or carry out underpinning works to a building within 3 metres of the centreline of a drain, sewer or disposal main to which the *Building (Amendment) Regulations 2001 (SI 2001/3335), Schedule 1, paragraph H4* applies, the Approved Inspector must consult the sewerage undertaker. The procedures and time periods involved parallel those described for fire authority consultations above.

5.8 Control over works in progress

5.8.1 Site inspections

The law is unclear regarding the extent to which an Approved Inspector needs to inspect works on site. Regulation 15 of the Building Regulations 2000 which governs the giving of notice for certain stages of work (see Section 5.4.2) does not apply where the work is supervised by an Approved Inspector. The functions which an Approved Inspector must carry out are specified and detailed in regulation 11 of the *Building (Approved Inspectors etc) Regulations 2000 (SI 20002532)*. His or her obligation is to '*take such steps (which may include the making of tests of building work and the taking of samples of material) as are reasonable to enable him to be satisfied within the limits of professional care and skill*' that the Building Regulations are complied with. Since this includes (in the case of dwellings) the need to receive the notice of energy rating and (in the case of dwelling-houses, flats and rooms used for residential purposes) the need to receive the notice of sound insulation testing, it would be rather difficult to carry out his or her function without site inspection.

Therefore, by virtue of regulation 11, an Approved Inspector is liable for negligence and it is likely that he or she is obliged to inspect the work to ensure compliance, unlike local authorities where there is discretion as to whether or not to inspect.

In any case, the Approved Inspector is contractually bound to the client, and the conditions of contract will usually include the need for certain inspections to be carried out of the work in progress. Failure to carry out such inspections will amount to breach of contract, with the usual remedies being available for such breaches.

Approved Inspectors who have embraced the *Building Control Performance Standards* are required to adopt an appropriate site inspection regime, which takes full account of relevant factors such as:

- the degree of detail in the design assessment process
- the nature of the work
- the experience of the builder
- the complexity and rate of building
- any unusual or high-risk features
- the arrangements for notifying that work is ready for inspection
- the key stages of construction.

These factors must be assessed at the outset, and regularly reviewed, so that effective control can be maintained for the duration of each project. It is also essential that adequate records are kept, which are sufficient to demonstrate the application of reasonable skill and care.

This means that records of each inspection should be maintained, which identify the work inspected and highlight any non-compliance. Details of non-compliant work must be communicated promptly and clearly to the responsible person, including an indication of any measures believed to be necessary for rectification of the situation. Additionally, mechanisms for appealing against or disputing the decision of an Approved Inspector should clearly be made known to the responsible person.

During the inspection phase, the Approved Inspector must ensure that all statutory consultees (e.g. the fire authority and the sewerage undertaker) are notified of any significant departure from the plans.

5.8.2 Testing of building work and sampling of materials

For Approved Inspectors, the powers to test building work are contained in regulation 11(1) of the *Building (Approved Inspectors etc) Regulations 2000* (as amended by the *Building (Approved Inspectors etc)(Amendment) Regulations 2001*).

Under regulation 11(1) Functions of Approved Inspectors '*an approved inspector by whom an initial notice has been given shall, so long as the notice continues in force, take such steps (which may include the making of tests of building work and the taking of samples of material) as are reasonable to enable him to be satisfied within the limits of professional skill and care that – (a) regulations 4, 6 and 7 of the Principal Regulations are complied with …*'

Regulation 11 appears to give the same powers to approved inspectors as regulation 18 gives to local authorities (see Section 5.4.9). At first reading the wording of regulation 11(1) seems to place the onus on the approved inspector to make the tests of building work (*an approved inspector … shall, … take such steps (which may include the making of tests of building work and the taking of samples of material) as are reasonable to enable him to be satisfied, etc….*); however the steps to be taken could include requiring another party (the contractor, etc.) to carry out the tests.

If it is accepted that approved inspectors may have the powers to require tests to be carried out by others but Local Authorities do not, this could result in unfair competition between local authorities and Approved Inspectors. For example, contractors could refuse to carry out air-leakage tests on buildings when using the local authority for building control (and therefore not have to pay for such tests) but be compelled to carry out tests when using the Approved Inspector system (and have to cover the costs).

In Section 5.4.9, it has been argued that regulation 18 does not permit a local authority to require a contractor (or any other person) to carry out an air-leakage test on a building. Such a requirement does exist in section 33 of the *Building Act 1984* but has never been activated (although section 33 does not apply to approved inspectors in any case). Until it is, it is difficult to see how the recommendations for air-leakage testing can ever be enforced by local authorities.

Sound insulation testing

The difficulties described above concerning air-leakage testing will not arise in the case of sound insulation testing, since by virtue of regulation 12A introduced by the *Building (Approved Inspectors etc) (Amendment) Regulations 2002*, an obligation is placed on the Approved Inspector (in the case of dwelling-houses, flats and rooms used for residential purposes) to check that appropriate sound insulation testing has been carried out by the developer to ensure compliance with regulation E1 (*Protection against sound from other parts of the building and adjoining buildings*). The results of the test must be recorded in a manner approved by the Secretary of State (see paragraph 1.41 of Approved Document E) and must be given to the Approved Inspector not later than 5 days after completion of the works.

The requirement for pre-completion sound insulation testing has been the subject of much debate, and the response to the consultation on Part E in 2002 showed that some sections of the building industry would prefer to develop robust standard details making testing of sound insulation for a sample of new homes on each site unnecessary. Therefore, the requirement for the Approved Inspector to be given details of the pre-completion sound insulation testing was introduced in stages to give the House Builders Federation time to gear up for the changes. Working with ODPM officials, the Federation developed robust standard details and these were the subjects of a consultation exercise by the ODPM in the latter part of 2003. Therefore, the situation is that with the exception of the requirements for pre-completion sound insulation testing of new houses and buildings containing flats, the requirements of the new Part E came into force on 1 July 2003. For new houses and buildings containing flats pre-completion testing may be done and submitted as evidence of compliance; however, an alternative has existed since 1 July 2004. From that date the requirement for testing does not apply where the person carrying out the building work notifies the approved inspector in accordance with the regulations that one or more of the design details published by Robust Details Ltd and specified in the notification has been adopted.

5.8.3 Non-conforming work and enforcement actions

Approved Inspectors are not empowered to enforce the Building Regulations (this can only be done by local authorities), however there are powers in Part VI of the *Building (Approved Inspectors etc) Regulations 2000 (SI 2000/2532)*, which can result in the cancellation of the initial notice and reversion of the work to the local authority, resulting in the resurrection of their full enforcement powers.

Reversion of the work to the local authority for enforcement is a final option rarely, if ever, used in practice. It will only arise where the Approved Inspector has tried all other courses of action without success, and feels that the extent of the contravention is so serious that the threat of cancellation of the initial notice is the only option left. In such circumstances, he or she may give a notice of contravention to the client specifying:

- those requirements of the Building Regulations which he or she feels are not being complied with
- the location of the non-compliant work.

The notice will inform the client that unless the contravention is rectified within a period of 3 months from its date of service, the initial notice will be cancelled. This gives the client a period of grace in which to instruct the builder to take down and remove the contravening work or to carry out alterations to make it comply.

Where the contravening works have not been rectified within the 3-month period the Approved Inspector will cancel the initial notice. This is done by serving a notice of cancellation on the local authority and the client (see Form 6 of the *Building (Approved Inspectors etc) Regulations 2000* as illustrated in Figure 6.1). This notice must give details of the contravention, unless a further initial notice relating to the work has been given and accepted. Full details of enforcement powers are given in Chapter 6.

5.8.4 The Final Certificate

When the Approved Inspector is satisfied that the project has been completed, he or she must give a '**Final Certificate**' to the Local Authority. A typical Final Certificate is shown in Figure 5.6. The local authority has 10 days in which to accept or reject the certificate, but it can only be rejected on the following prescribed grounds:

- The certificate is not in the prescribed form.
- It does not describe the work to which it relates.
- No initial notice relating to the work is in force.
- The certificate is not signed by the Approved Inspector who gave the initial notice or that he or she is no longer an Approved Inspector.
- Evidence of approved insurance is not supplied.
- There is no declaration of independence (except for minor work).

If the local authority rejects a final certificate, the initial notice to which it refers will cease to be in force within 4 weeks of the date of rejection.

Acceptance of the final certificate effectively removes the local authority's powers to take proceedings against the client for a contravention of Building Regulations in relation to the work referred to in the final certificate. The Approved Inspector does not have to copy the final certificate to the any other party, but it is usual for a certificate of completion to be issued to the client once the local authority has accepted the final certificate.

FORM 5

Section 51 of the Building Act 1984 ('the Act')

The Building (Approved Inspectors etc) Regulations 2000 ('the 2000 Regulations')

FINAL CERTIFICATE

1 This certificate relates to the following work (Note 1)

....

....

....

2 I am an approved inspector for the purposes of Part II of the Act and the above work is [the whole]/[part](Note 2) of the work described in an initial notice given by me and dated.(Note 3)

3 Subject to what is said in paragraph 4 below, the work described above has been completed and I have performed the functions assigned to me by regulation 11 of the 2000 Regulations.

4 The work described above involves the insertion of insulating material into a cavity wall and this [has]/[has not](Note 2) been carried out.(Note 4)

5 Final certificates have now been issued in respect of all work described in the initial notice referred to in paragraph 2 above.(Note 4)

6 With this certificate is the declaration, signed by the insurer, that a named scheme of insurance approved by the Secretary of State applies in relation to the work to which the certificate relates.

7 The work [is]/[is not](Note 2) minor work.(Note 5)

8 I have had no financial or professional interest in the work described since giving the initial notice described in paragraph 2 above.(Note 4)

Signed...

Approved Inspector

Date........................

NOTES

(1) Location and description of the work including the use of any building to which the work relates.
(2) Delete whichever does not apply.
(3) Insert date.
(4) Delete this statement if it does not apply.
(5) 'Minor work' has the meaning given in regulation 10(1) of the 2000 Regulations. If work is not minor work, the declaration in paragraph 8 must be made.

Figure 5.6 Typical final certificate.

The client should take great care if it is proposed to occupy a building (or part of a building) before completion. If no final certificate has been given before occupation then the initial notice will cease to have effect on the expiry of certain time periods from the date of occupation, and the work will revert to the local authority. Thus, for most buildings there is a period of grace of 8 weeks from the date of occupation, before the initial notice lapses. However, if the work involves the erection, extension or material alteration of a building, and it is to be put to a relevant use as a workplace of a kind to which the *Fire Precautions (Workplace) Regulations 1997, Part II* applies or a use designated under the *Fire Precautions Act 1971, section 1*) the period is reduced to 4 weeks. Currently designated are hotels, boarding houses, offices, shops, railway premises and factories.

If the client needs to occupy part of a building whilst work is continued on the remainder a final certificate may be given for the part to be occupied and further final certificates given as future phases of the work are completed and/or occupied. In this way it is possible to avoid the lapse of the initial notice and the reversion of the work to the local authority. It is of course essential that, at the outset, the client makes the need for such phasing of occupation and completion clear to the Approved Inspector, to avoid later misunderstandings.

Once the initial notice has ceased to have effect, the Approved Inspector will be unable to give a Final Certificate and the Local Authority's powers to enforce the Building Regulations can revive unless another Approved Inspector is engaged by the client. If the Local Authority becomes responsible for enforcing the Regulations it must be provided on request with plans of the building work so far carried out. Additionally, it may require the person carrying out the work to cut into, lay open or pull down work so that it may ascertain whether any work not covered by a final certificate contravenes the Regulations. If it is intended to continue with partially completed work, the local authority must be given sufficient plans to show that the work can be completed without contravention of the Building Regulations. A fee, which is appropriate to that work, will be payable to the local authority.

5.8.5 Conclusions

Thus, in the normal course of events, there should be a seamless interaction between the Building Regulation assessment service and the normal design, and construction process provided that adequate channels of communication are established and maintained. Questions of compliance can be sorted out in good time (either during the design process or on site) before they become a problem with potential to disrupt the construction programme or add to the construction costs. Therefore, the inspection framework should ensure that, in most cases, the building work will continue to a successful conclusion, and will be in conformity with the Building Regulations.

Experience shows that the vast majority of queries over compliance can be dealt with in this way without involving the necessity for legal action of any kind. Approved Inspectors are not empowered to enforce the Building Regulations (this can only be done by local authorities), however there are powers in the *Building (Approved Inspectors etc) Regulations 2000 (SI 2000/2532)*, which can result in the cancellation of the initial notice and reversion of the work to the local authority, resulting in the resurrection of their full enforcement powers.

Reversion of the work to the local authority for enforcement is a final option rarely, if ever, used in practice. It will only arise where the Approved Inspector has tried all other courses of action without success, and feels that the extent of the contravention is so serious that the threat of cancellation of the initial notice is the only option left.

BUILDING CONTROL BY PUBLIC BODIES

5.9 Introduction

Part VII of the *Building (Approved Inspectors etc) Regulations 2000* is concerned with public bodies and, read in conjunction with section 54 and Schedule 4 of the *Building Act 2000*, its effect is to enable designated public bodies to self-certify their own work to which the substantive building regulations apply.

Public bodies are approved by the Secretary of State (or in Wales, the National Assembly of Wales), and the regulations relating to notices, consultation with the fire authority and the sewerage undertaker, plans certificates and final certificates mirror those of Part III of the *Building Act* 1984 dealing with Approved Inspectors. The grounds on which the local authority may reject a public body's notice, etc., mirror those applicable to the approved inspector system, except that:

- there is no provision for cancellation of a public body's notice
- there is no requirement that there should be an approved insurance scheme in force
- unlike approved inspectors, public body's do not have to be independent of the work they are supervising under the regulations (i.e. they can carry out a building control service on projects which they have designed and are building themselves.

To date, only the Metropolitan Police Authority has been prescribed as a public body under the provisions of the *Building Act 1984, section 5*.

5.9.1 Public body's notice

Where a designated public body considers that it can adequately supervise its own building work using its own staff it may serve a public body's notice on the local authority in whose area the work is located. A typical public body's notice is shown in Figure 5.7 which is based on Form 9 from Schedule 2 of the 2000 Regulations.

The local authority can reject the notice only if it is defective in terms of accuracy and completeness. The grounds for rejection are laid out in Schedule 6 to the 2000 Regulations as follows:

- The notice is not in the prescribed form.
- The notice has been served on the wrong local authority.
- The body on behalf of which the notice was signed is not a public body within the meaning of section 54 of the Act.

A.N. Other public body

Public body's notice

Section 54 of the Building Act 1984 ('the Act')
The Building (Approved Inspectors etc) Regulations 2000 ('the 2000 Regulations')
FORM 9

To: (Note 1)

....
....
....
....

---

1. This notice relates to the following work: (Note 2)

Location: ....
....

Description of work: ....

....

Use of any building: ....

---

2. .... (Note 3) is approved under Part II of the Act and intends to carry out in relation to a building belonging to it the work described above which can be adequately supervised by its own servants or agents.

---

3. With this notice are the following documents, which are those relevant to the work described in this notice: (Note 4)

(a) in the case of the erection or extension of a building, a plan to a scale of not less than 1:1250 showing the boundaries and location of the site, and where the work includes the construction of a new drain or private sewer, a statement:
(i) as to the approximate location of any proposed connection to be made to a sewer, or
(ii) if no connection is to be made to a sewer, as to the proposals for the discharge of the proposed drain or private sewer including the location of any septic tank and associated secondary treatment system, or any wastewater treatment system or any cesspool.

(b) the following **Local Enactment** is relevant to the work (*delete if none applies*): -

....

The following steps will be taken to comply with it:-

....
....

A.N. Other Public Body declares that:

(i) It will/will not* be obliged to consult the fire authority by regulation 23 of the 2000 Regulations.(Note 5)

(ii) It undertakes to consult the fire authority before giving a public body's plans certificate in accordance with paragraph 2 of Schedule 4 to the Act, or a public body's final certificate in accordance with paragraph 3 of Schedule 4 to the Act in respect of any work described above.(Note 6)

Figure 5.7 Typical public body's notice.

(iii) It will/will not* be obliged, by regulation 23A of the 2000 Regulations, to consult the sewerage undertaker. (*If the sewerage undertaker has to be consulted the next following declaration must be made*)

(iv) It undertakes that before giving a public body's plans certificate in accordance with paragraph 2 of Schedule 4 to the Act, or a public body's final certificate in accordance with paragraph 3 of Schedule 4 to the Act, it will consult the sewerage undertaker in respect of any of the work described above. (Note 6)

(*delete whichever statement is inapplicable)

Signed...   Name..

For and on behalf of A.N. Other Public Body

Date..

Notes

(1) Insert the name and address of the local authority in whose area the work will be carried out.
(2) Location and description of the work and the use of any building to which the work relates.
(3) Insert the name and address of the public body.
(4) The local authority may reject this notice only on grounds prescribed by the Secretary of State. These are set out in Schedule 6 to the 2000 Regulations. They include failure to provide the relevant documents. The documents listed in this section of the notice relevant to the work described above should therefore be sent with this notice. Any sub-paragraph which does not apply should be deleted.
(5) If the inspector is obliged to consult the fire authority, the next following declaration **must** be made.
(6) Delete this statement if it does not apply.

Figure 5.7 (Continued).

- The information supplied is deficient because neither the notice nor plans show the location or contain a description of the work (including the use of any building to which the work relates).
- The public body is obliged to consult the sewerage undertaker before giving a public body's plans certificate or public body's final certificate and the notice does not contain an undertaking to do so.
- Where it is intended to erect or extend a building, the local authority considers that a proposed drain must connect to an existing sewer, but no such arrangement is indicated in the notice.
- The notice does not contain an undertaking to consult the fire authority (where this is appropriate).
- Local legislative requirements will not be complied with.

The events that can cause a public body's notice to cease to be in force are identical to those which apply to an initial notice under the approved inspector system and are covered by regulation 18 of the 2000 Regulations (see Section 5.6.5).

A public body's notice may be combined with a public body's plans certificate. There is a prescribed form for this (see Form 11 in Schedule 2 of the *Building (Approved Inspectors etc) Regulations 2000*).

5.9.2 Public body's plans certificate

A public body may give a plans certificate in a similar manner to that for an approved inspector. It may also be combined with a public body's notice, at the discretion of the public body. Prescribed forms (Form 10 for a plans certificate and Form 11 for a combined certificate) are presented in Schedule 2 to the 2000 Regulations, and the grounds for rejection are given in Schedules 6 and 7. The local authority must reject the public body's notice, combined notice and plans certificate, or plans certificate as appropriate, within 10 days of the notice being given or it is deemed to have been accepted. The effect of the plans certificate is similar to that for an approved inspector's plans certificate described in Section 5.6.7.

5.9.3 Public body's final certificate

The prescribed form for a public body's final certificate is presented in the Schedule 2 of the 2000 Regulations as Form 12, and the grounds for its rejection are given in Schedule 8. The certificate is similar to an approved inspector's final certificate (see Figure 5.6) except that it does not contain a Declaration of independence (and therefore the need to reveal the existence of minor work) or a Declaration that an approved insurance scheme is operative (since none is needed for a public body). The grounds for rejection are also similar to those that apply to approved inspectors and are similarly truncated.

WORK UNDER THE SUPERVISION OF A COMPETENT PERSON

5.10 Background

In recent years the Government has been keen to extend control over work which can affect the energy efficiency of buildings. In April 2002 new requirements came into force bringing under control replacement windows, rooflights, roof windows and doors, and hot water vessels. At the same time the Government was conscious of the fact that the new requirements would increase the administrative burden on local authorities and Approved Inspectors for what was, in fact, fairly minor work. As a result it was thought appropriate to introduce new ways of controlling such work by means of 'self-certification' schemes and 'non-notification'. The essence of such systems is that they are self-policing. The difference between them is that self-certification schemes require the person actually carrying out the work to notify the local authority on completion, whereas for non-notification no such requirement exists. One such non-notification system has been in existence for many years whereby a gas appliance could be installed by a person, or an employee of a person, approved in accordance with regulation 3 of the *Gas Safety (Installation and Use) Regulations*. This non-notification system has now been extended to a range of building operations covered by the regulations and described more fully below. The first self-certification scheme, the Fenestration Self-Assessment Scheme (FENSA) came into force on 1 April 2002.

5.11 Self-Assessment

5.11.1 Self-certification schemes

Regulation 16A of the *Building Regulations 2000* (as amended) allows local authorities to accept certificates given by certain persons, as evidence that the requirements of regulations 4 and 7 have been met in relation to work described in Schedule 2A and discussed below. Such individuals must be suitably qualified and experienced and/or registered persons under a recognised scheme as described in Schedule 2A. Whether the local authority is prepared to accept the certificate is optional. However, if the authority does not accept it and decides to inspect the work, it may incur certain obligations and liabilities attendant upon exercising its building control function.

Under regulation 16A local authorities must either be informed within 30 days of completion of the work by the person carrying it out, or must be given the certificate referred to above. A copy of the certificate must also be given to the occupier of the building where the work is carried out.

The need to notify completion or give a certificate to the local authority and the occupier does not apply to building work described in Schedule 2B (see below) which consists only of work on a low-voltage or extra-low-voltage electrical installation.

5.11.2 The FENSA scheme

To assist the effective implementation of the recent amendments to Part L (Conservation of Fuel and Power) a scheme for the self-certification of replacement glazing (the Fenestration Self-Assessment Scheme, FENSA) was set up by the Glass and Glazing Federation. In order to take part in the scheme, persons (companies or individuals) must be registered with FENSA Ltd. Currently, the FENSA scheme limits registration of installers to replacement windows, rooflights, roof windows and doors in dwellings only, subject to the work not having structural implications for a building.

5.11.3 Operation of the FENSA scheme

Following completion of the replacement window installation in a dwelling, installers who are self-certifying their work send the certificate to the FENSA database. The system then consolidates all certificates and forwards them to the appropriate local authority within 5 days of receipt from the installer. Paper certificates are also prepared by FENSA for forwarding to the appropriate householder. This is important, since evidence of compliance with the regulations will need to be supplied to any future purchaser of the house.

In order to operate the scheme FENSA maintains a database of all its members and all replacement window installations that have been certified by FENSA registered companies. The database system is used to:

- receive authenticated transaction data from installation companies
- forward certificates to local authorities
- receive membership data from FENSA
- receive and provide inspection data from/to FENSA inspectors
- prepare and despatch paper certificates to householders
- permit secure interrogation for authorised persons.

The preferred method of forwarding certificates to local authorities is electronically; however, where a local authority cannot receive electronic certificates these are sent by facsimile. Additionally, FENSA can allow access to the database so that local authorities can obtain information about individual company registrations and particular properties within their area, subject to security authentication procedures that confirm the identity of the enquirer.

Independent checks of installation companies are carried out on an activity-sampling basis by FENSA inspectors in order to ensure that the regulation requirements are being met.

Where replacement windows are installed by companies or individuals not registered under the FENSA scheme the work will be subject to the normal building control procedures whereby the person carrying out the work will need to give a building notice or submit full plans to a local authority, or give an initial notice jointly with an Approved Inspector.

5.11.4 Combustion appliances and oil-storage facilities

For the following types of installation the person carrying out the work is not required to give a building notice or deposit full plans provided that they are suitably qualified and experienced and/or registered under a recognised scheme as mentioned below, and they notify the local authority of completion or give a certificate as mentioned in Section 5.11.1. In general the option to undertake appliance installation work by self-certification is limited to work within buildings of no more than three storeys (not including basements). However, for gas appliance installers, there is no such limitation on installation of the appliance itself but the self-certification option is so restricted for associated work.

Installation of a heat-producing gas appliance

The term 'appliance' includes any fittings or services (other than an unvented hot water storage vessel), which form part of the space heating or hot water system served by the combustion appliance. The person (or an employee of that person) carrying out the work must be approved in accordance with regulation 3 of the *Gas Safety (Installation and Use) Regulations 1998*. The regulations are concerned with controlling the risks which arise when using gas from either mains pipes or gas storage vessels. The work must be carried out by a competent person who is a member (or is employed by a member) of the Council for Registered Gas Installers (CORGI).

Installation of oil-fired combustion appliances and oil storage tanks

The self-certification option applies only to oil-fired combustion appliances with a rated heat output of 45 kilowatt or less, and to oil storage tanks, installed by individuals registered under the Oil Firing Registration Scheme by the Oil Firing Technical Association for the Petroleum Industry (OFTEC) for that type of work. The operatives concerned must be registered with OFTEC as:

(a) installation technicians and domestic/light commercial commissioning and service technicians for pressure jet and for vaporising burners
(b) commercial industrial commissioning and service technicians
(c) oil storage tank installation technicians.

Although the self-certification option is limited to registered individuals, others who are not registered often assist installation. In these cases, supervision by the registered person is the same as installation by that person.

Installation of solid fuel burning combustion appliances

The self-certification option applies only to solid fuel burning combustion appliances with a rated heat output of 50 kilowatt or less, installed by individuals registered with the Registration Scheme for Companies and Engineers involved in the Installation and Maintenance of Domestic Solid Fuel Fired Equipment operated by HETAS Ltd for that type of work. The operatives concerned are those registered for the installation and commissioning of domestic central heating/ hot water systems, and the installation and commissioning of domestic appliances not connected to a hot water system. Although the non-notification option is limited to registered individuals, others who are not registered often assist installation. In these cases, supervision by the registered person is the same as installation by that person.

Installation and commissioning

Whereas for gas and solid fuel the installer also commissions or brings into use the appliance, this is not always the case for oil-fired combustion appliances. For this reason individuals are required to register separately as installers and commissioning technicians. They may be the same person or two different individuals; however, both aspects of the work (initial installation and commissioning) must have been completed before the installation can be regarded as complete.

5.11.5 Other building work

The self-certification option extends to any building work which is necessary to ensure that any appliance, service or fitting which is installed, and which is described above, complies with the applicable requirements contained in Schedule 1 of the Building Regulations 2000. The person carrying out the building work would be the person who installed the appliance, service or fitting referred to in the sections above.

For example, where a combustion appliance was being installed, this could mean the provision of flue pipes, and work required to ensure compliance with Part L, of a space heating or hot water system served by the combustion appliance. In the case of the installation of cast *in situ* flue linings, the material and installation procedures would need to be independently certified. Even if he or she does not actually carry out the building work himself or herself, the installer of the combustion appliance must supervise the work, and through that supervision, take responsibility for it.

The building work referred to in this section which is necessary to ensure that a gas combustion appliance complies with the Schedule 1 of the Building Regulations can only be carried out by the person referred to in Section 5.11.4, if the appliance has a net rated heat input of 70 kilowatt or less and is installed in a building of three storeys or less (not counting basements).

The self-certification option does not extend, however, to the provision of a masonry chimney.

5.11.6 Installation of fixed low- or extra-low voltage electrical installations

Part P of the *Building Regulations 2000* (as amended by the *Building (Amendment) (No 2) Regulations 2004*) applies to electrical installations that are intended to operate at low or extra-low voltage (see regulation 2 in Appendix 1 for definitions). Such installations are generally associated with dwellings (including common parts in flats, sheds and greenhouses, and garden lighting, etc.) and Part P is restricted to such uses. In general, the self-certification option is available as described above for installations covered by Part P. However, this is further extended to a 'non-notification' system of control for the following types of work described in Schedule 2B to the *Building Regulations 2000* (as amended):

- Work consisting of:
  - (a) replacing any socket-outlet, control switch or ceiling rose
  - (b) replacing a damaged cable for a single circuit only
  - (c) re-fixing or replacing enclosures or existing installation components, where the circuit protective measures are unaffected
  - (d) providing mechanical protection to an existing fixed installation, where the circuit protective measures and current carrying capacity of the conductors are unaffected by the increased thermal insulation.
- Work which:
  - (a) is not in a kitchen, or a special location
  - (b) does not involve work on a special installation
  - (c) consists of:
    - (i) adding light fittings and switches to an existing circuit
    - (ii) adding socket-outlets and fused spurs to an existing ring or radial circuit or
    - (iii) installing or upgrading main or supplementary equipotential bonding.

'Special installation' above means an electric floor or ceiling heating system, a garden lighting or electric power installation, an electricity generator, or an extra-low voltage lighting system which is not a pre-assembled lighting set bearing the CE marking referred to in regulation 9 of the *Electrical Equipment (Safety) Regulations 1994*; and 'special location' means a location within the limits of the relevant zones specified for a bath, a shower, a swimming or paddling pool or a hot air sauna in the *Wiring Regulations, 16th edition*, published by the Institution of Electrical Engineers and the British Standards Institution as *BS 7671: 2001* and incorporating amendments 1 and 2.

Therefore, all work to electrical installations in or associated with dwellings must now comply with Part P of the *Building Regulations 2000* (as amended by the *Building (Amendment) (No 2) Regulations 2004*) and as such must be carried out by a person registered by one of the following bodies:

- BRE Certification Ltd,
- British Standards Institution,
- ELECSA Ltd,
- NICEIC Certification Services Ltd,
- Zurich Certification Ltd,

for that type of work, unless it is a type of work described in Schedule 2B.

Where the work falls outside the defined restrictions given in Schedule 2B it is then subject to self-certification as described in Section 5.11.1.

6

Enforcement actions

6.1 Contravening works

Local authorities have extensive powers under the provisions of the *Building Act 1984* to ensure that building work covered by the Building Regulations does in fact comply with those regulations. As has been explained in Chapter 5, these enforcement powers even extend to work being supervised by an Approved Inspector, under certain conditions.

6.1.1 Functions and duties of local authorities to enforce the regulations

Section 91(2) of the *Building Act 1984* states that it is the function of local authorities to enforce Building Regulations in their areas. Since section 91 deals with 'duties' of local authorities it would appear that the function to enforce is also a duty, in which case a local authority does not have discretion to decide not to take steps to enforce once it has decided that there is a *prima facie* contravention and that action is needed to ensure compliance.

6.1.2 Powers of entry

Under section 95 of the *Building Act 1984* an authorised officer of the local authority has a right to enter premises for the purposes of:

- ascertaining whether there is, or has been, a contravention of the regulations
- ascertaining whether circumstances exist which would authorise the local authority to take any action or execute any work under the regulations
- taking any action or executing any work authorised by the regulations
- performance of the local authority's functions under the regulations.

The authorised officer must produce some duly authenticated document showing his or her authority. Entry may be made at all reasonable hours and, except in the case of a factory or workplace, entry may not be demanded as of right unless 24 hours' notice has been given to the owner or occupier.

Where exceptional circumstances exist, such as:

- admission has been refused, or a refusal is apprehended
- the premises are unoccupied
- the occupier is temporarily absent
- the case is urgent
- an application for admission would defeat the object of entry

and there is reasonable ground for entry into the premises for any of the purposes listed above, an application may be made to a justice of the peace for a warrant authorising entry, if necessary by force. Section 96 of the *Building Act 1984* requires that the duly authorised officer must ensure that on leaving, unoccupied premises are adequately secured.

6.1.3 Causes of contravention

When using a local authority the possible causes of contravention are:

- Failure to give a building notice or submit full plans for controlled work.
- Failure to serve a notice of commencement on the local authority before starting work.
- Starting work after service of a commencement notice but before 2 days have elapsed since its service.
- Covering up the following work without notifying the local authority and/or covering up the work without allowing the local authority at least one days' notice to inspect (the notice period runs from the end of the day on which the notice was given) any:
  - excavation for a foundation
  - foundation
  - damp-proof course
  - concrete or other material laid over a site
  - drain or sewer to which the regulations apply.
- Failing to notify the local authority within 5 days of the following being completed:
  - when a drain or sewer which is subject to the requirements of Part H (Drainage and waste disposal) has been laid, haunched or covered up
  - the building work which is the subject of the building notice or full plans submission.
- Failing to notify a local authority at least 5 days before a building (or part) is to be occupied if this is to take place before completion.
- Carrying out work which is not in compliance with Schedule 1 of the Regulations.

When using an Approved Inspector the only possible cause of contravention is non-compliance with Schedule 1. The Approved Inspector cannot take direct action but in extreme circumstances (e.g. where the Approved Inspector is unable to get his or her client to put right a case of non-compliance) he or she must refer the contravention back to the relevant Local Authority for enforcement action.

6.1.4 Section 36 notice

Under section 36 of the *Building Act 1984*, where a building is erected, or building work is done contrary to the regulations, the local authority may serve a notice on the building owner requiring him or her to:

- pull down or remove the offending work or
- alter the work so that it complies with the regulations.

A person who has carried out any further work to secure compliance with the regulations must give notice to the local authority of its completion, within a reasonable time.

Where the work is required to be removed or altered, and the owner fails to comply with the local authority's notice within 28 days (or such longer time as may be allowed by a magistrate's court on application by the owner), the local authority may remove the contravening work or execute the necessary work itself so as to ensure compliance with the regulations, recovering expenses in so doing from the defaulter.

A section 36 notice cannot be given:

- after the expiration of 12 months from the date on which the work was completed or
- where the local authority has passed the plans (even though the plans were defective) and the work has been carried out in accordance with those plans or
- notice of rejection was not given within the relevant time period from the deposit of plans and the work has been carried out in accordance with them.

The 12-month time limit specified by the Building Act is in practice reduced to 6 months by virtue of the provisions of section 127(1) of the *Magistrate's Court Act 1980* since it has been held that the person constructing a building commits an offence when the building works are completed in a way not complying with the regulations. He or she does not commit a continuing offence. This means that proceedings must be commenced within 6 months of the commission of the alleged offence; that is, when the building works were completed.

6.1.5 Appeal against a section 36 notice

The recipient of a section 36 notice has a right of appeal to the magistrate's court under section 40 of the *Building Act 1984*. The burden of proving non-compliance with the regulations lies with the local authority, but if it can show that the works do not comply with an Approved Document (see Chapter 7) the burden of proof of compliance with the regulations shifts to the appellant against the section 36 notice (e.g. the building owner).

There is an alternative to the ordinary appeal procedure, under section 37 of the *Building Act 1984*. Under that section, the owner may notify the local authority of his or her intention to obtain from a 'suitably qualified person' a written report concerning work to which the section 36 notice relates. Such notices are usually served where the local authority considers that the technical requirements of the regulations have been infringed.

The report from the suitably qualified person is then submitted to the local authority, which may withdraw the section 36 notice, after consideration of the contents of the report. Section 37 also makes provision for the local authority to pay to the building owner the expenses that it considers that he or she has reasonably incurred in consequence of the service of the notice, including his or her expenses in obtaining the report. Adopting this procedure has the effect of extending the time for compliance with the notice or appeal against it from 28 to 70 days.

Experience shows that in most cases the local authority will reject the findings of the report. In this case the report can be used as evidence in any appeal under section 40 of the *Building Act 1984* and section 40(6) provides that:

> *If, on an appeal under this section, there is produced to the court a report that has been submitted to the local authority under section 37(1) above, the court, in making an order as to costs, may treat the expenses incurred in obtaining the report as expenses incurred for the purposes of the appeal.*

Thus, in the normal course of events, if the appeal was successful, the owner would recover the cost of obtaining the report as well as his or her other costs.

If a person is aggrieved by the decision of a magistrate's court he or she may appeal to the Crown Court (section 41, *Building Act 1984*).

6.1.6 Applying for an injunction

The local authority, or anyone else, may also apply to the civil courts for an injunction requiring the removal or alteration of any contravening works (section 36(6), *Building Act 1984*). This power is exercisable even in respect of work which has been carried out in accordance with deposited plans; for example, where there has been an oversight or mistake on the part of the local authority. In such a case the court is empowered to order the local authority to pay compensation to the owner. The 12-month time limit does not apply to this procedure which is highly unusual and rarely invoked in practice; however, it does provide a course of action if a building was considered to be a serious danger to the public. There are other procedures under the *Building Act 1984*; however, that are more usually used to control dangerous building structures. The Attorney-General, as guardian of public rights, may seek an injunction in similar circumstances, and in practice proceedings for an injunction must be taken in his or her name and with the consent.

6.1.7 Fines for contravention

Where a person contravenes any provision in the Building Regulations, the person renders himself or herself liable to prosecution by the local authority and the case is dealt with in the magistrates' court. The maximum fine on conviction is level 5 on the standard scale (see Section 3.2.4) with a further fine not exceeding £50 for each day on which the default continues after he or she is convicted.

6.2 Controlling previously completed work

6.2.1 Applying for a regularisation certificate

As has been shown above, local authorities have powers under section 36 of the *Building Act 1984* to take action against the owner of building work if it is carried out without a building notice being given or without the deposit of full plans. This does not apply, of course, if the owner has used an Approved Inspector or the work is subject to a public body's notice. Even where a building notice has been deposited or a full plans submission has been made the work will still be regarded as 'unauthorised' building work if no commencement notice has been given to the Local Authority.

Alternatively, if the unauthorised work was commenced on or after 11 November 1985 the owner may apply to the local authority for a 'regularisation certificate'. Although there may be no legal imperative for the application (e.g. where the unauthorised work has been completed for more than 12 months), the procedure is often used where the building is to be sold and the local authority search reveals that approval was not sought for the development. This can hold up or prevent the sale.

Regulation 21 of the *Building Regulations 2000* (as amended) sets out the procedures to be adopted where this course of action is decided upon by the building owner.

The application for a regularisation certificate must be in writing and must include:

(a) a statement that the application is made in accordance with regulation 21
(b) a description of the unauthorised work
(c) so far as is reasonably practicable, a plan of the unauthorised work
(d) so far as is reasonably practicable, plans showing any additional work which is needed to ensure that the unauthorised building work complies with the regulations. (*Note*: The work will need to comply with the regulations which were in force at the time that the unauthorised work was carried out, therefore this date may need to be proved to the local authority.)

When the local authority receives an application for a regularisation certificate it may require the owner to lay open the work or make tests and provide samples of materials in order that it can decide what work, if any, is needed to ensure compliance with the regulations.

When the local authority is satisfied that it has sufficient information it will notify the owner of the work that is needed (if any) in order to satisfy the relevant requirements of the Regulations. If a relaxation or dispensation of the Regulations (Section 6.3.3) has been approved the local authority must take this into account when assessing any work that is needed.

If the local authority is satisfied that the work complies with the Regulations (i.e. those that were in force when the original unauthorised work was carried out) it may give the owner a regularisation certificate.

An application for a regularisation certificate and the plans which accompany it are not regarded as the deposit of plans under section 16 of the *Building Act 1984*.

6.3 Dealing with questions of compliance

6.3.1 Introduction

The interpretation and administration of the building regulations is a complex business and it is inevitable that, at some stage, disagreements will arise between the person carrying out the building work and the building control body (Local Authority or Approved Inspector) with regard to application and methods of compliance.

Accordingly, the *Building Act 1984* contains a number of provisions aimed at addressing the situations where disputes arise by providing various resolution pathways.

The Building Regulations impose mandatory requirements on people who propose to carry out building work with respect to the design and construction of buildings. The Regulations are specifically targeted at public health, safety, welfare and convenience, and also address issues of energy conservation.

They are phrased in functional terms and contain no practical guidance regarding methods of compliance. Non-mandatory guidance is provided principally by Approved Documents and Harmonised Standards (British or European); however, there are other methods of demonstrating compliance, such as past experience of successful use, test evidence, calculations, compliance with European Technical Approvals, the use of CE-marked materials, etc.

Approved Documents and Harmonised Standards are generally the most appropriate way to achieve compliance with the requirements of the Building Regulations; however, the Building Control Body must judge any relevant building proposals against the provisions of the Regulations, and not the guidance documentation. If the Approved Document guidance is not followed, at present the onus is on the person carrying out the building work to demonstrate by other means that the requirements have been satisfied. Methods of compliance are discussed completely in Chapter 7.

Therefore, problems may arise (especially if the Approved Document guidance is not followed) if either:

(a) The person carrying out the building work believes that the plans show compliance with the requirements of the Regulations but the Local Authority or Approved Inspector does not agree with their interpretation or
(b) The person carrying out the work accepts that the plans do not fully comply with the requirements of the Building Regulations, but feels that in their particular circumstances, the requirements are too onerous.

In case (a) it would be open to the aggrieved person to apply for a determination, whereas in case (b) an application for a relaxation or dispensation of the non-complying requirement(s) might be the appropriate course of action.

6.3.2 Determinations and appeals

Applying for a Determination

An essential prerequisite for a valid determination application is that the applicant accepts that the particular requirement of the Building Regulations which is in question does in fact apply to the building work, and believes that plans of the proposed work comply with it. It is important to realise that the procedure is a determination and not some form of 'appeal' procedure against any decision made by a building control body.

The Determination procedure enables the person carrying out the work to apply to the Secretary of State (or, if the work is in Wales, to the National Assembly for Wales (NAW)) to determine whether or not plans of the proposals comply with the Regulations (see the *Building Act 1984*), *section 16(10)(a)* if building control is being carried out by a Local Authority or *Building Act 1984, section 50(2)* if an Approved Inspector has been appointed.

The need for a Determination usually arises because the Local Authority rejects the plans or the Approved Inspector refuses to give a plans certificate on the grounds that the plans contravene some part of the Regulations. Since it is necessary to have 'plans' of the proposed work, the Determination procedure is not available where the Local Authority 'building notice' procedure has been used.

A Determination can be applied for at any time once plans have been submitted to the Local Authority or after an Approved Inspector has been asked to supply a plans certificate. However, the need will not usually arise until after the plans have been examined and the applicant has been told informally that they are not acceptable. Obviously, the sooner that an application for a Determination is made the less delay there will be to the building work. However, because the *Building Act, section 16(10)(a)* refers to 'plans of the *proposed* work' (and the *Building Act 1984, section 50(2)* refers to plans of the work supplied by a person *intending* to carry out the work) it is important that these provisos are not compromised and applications for Determination must usually be made before any work commences.

In exceptional circumstances valid applications may be made after work has commenced, provided that the work has not progressed to such an extent that it compromises any options necessary to achieve compliance of a particular element of that work, and that full plans approval or a plans certificate has not been given. This will be a matter for the applicant and the building control body to decide, if necessary in consultation with Office of the Deputy Prime Minister (ODPM) or the National Assembly for Wales (NAW). When work has reached a stage where it has patently ruled out all other options, it is clear that an application for Determination could not be described at this stage as one of 'plans of proposed work', and it would not be accepted.

A Determination application should include:

- the names and addresses of the parties involved, including any agents
- the details of the Local Authority or Approved Inspector providing the building control service
- the full address of where the proposed building work will be carried out
- a statement setting out details of the building, the proposed work and the matter in dispute
- a statement setting out the case for compliance with the particular Regulation requirement in question
- a copy of the plans of the proposed work and any other documents which have been submitted to the Local Authority with the full plans application, or those submitted to the Approved Inspector on which he or she was unable to give a plans certificate
- a copy of all relevant correspondence with the Local Authority or Approved Inspector involved, including the notice of rejection of plans if one has been issued
- a copy of any listed building consent if required for the proposed work and any associated planning permission relevant to the listed building
- a copy of any other documents supporting the case for compliance, including calculations
- where appropriate, a location or block plan and photographs of the proposed work to illustrate particular points
- the appropriate fee (i.e. half the local authority's plan charge, with a minimum of £50 and a maximum of £500 payable).

The application should be sent to the ODPM if the proposal is in England or the NAW if it is in Wales. As there is no model application form, the application should be in the form of a letter. The agreement of the Local Authority or Approved Inspector is not needed in order to ask for a Determination.

Once the application is accepted it will be acknowledged, and will be copied to the building control body for its comments. The building control body will be given 21 days to respond and its response will be copied to the applicant. The applicant has 21 days in which to respond to the comments of the building control body, by writing to the Secretary of State or NAW as appropriate. This response should be copied to the building control body. The Secretary of State or the NAW will give detailed consideration to the documentation submitted and will give the applicant and the building control body, a decision ('the Determination').

The Secretary of State or NAW does not make site visits. The determination application is considered on the basis of the written representations. If it is necessary to obtain further information from either party during the technical appraisal, the Secretary of State or NAW will seek this in writing and will copy the information received to the other party.

The Secretary of State or NAW aims to issue a decision within 3 months of the receipt of all documentation. Therefore, the whole process will take about 6 months on average with the most complex cases taking longer.

6.3.3 Applying for a relaxation or dispensation

An application for a relaxation or dispensation must be made to the relevant local authority. It will ask them to relax a requirement to some degree or to set aside (i.e. dispense with) a particular requirement altogether. This applies irrespective of whether an Approved Inspector or Local Authority is being used for building control services. If the local authority refuses the application an appeal can be made to the Secretary of State for the ODPM or the NAW (see below).

As explained in Section 6.3.1, the Building Regulations are phrased in functional terms with the intention of giving designers and builders flexibility in the way they comply. However, the consequence of this is that it may be very difficult for a local authority to selectively relax some aspects of a functional requirement which is written in terms of reasonable standards of health and safety, or other reasonable provisions. Additionally, if a decision is made to appeal against a refusal of a local authority to grant a relaxation or dispensation it is likely to be very difficult to argue the case unless very special circumstances exist.

An analysis of the appeals referred to the DTLR/ODPM will bear this out. Of the 31 appeals which were considered by the DTLR between February 1998 and November 2001 only one (reference 45/3/128) was decided in favour of the appellant and then only because it was held that the Regulation concerned (M2) did not apply in this case.

If a decision is taken to apply to the local authority for a relaxation or dispensation in accordance with section 8 of the *Building Act 1984*, the applicant should clearly state whether he or she is seeking either a relaxation or a dispensation, and should specify the precise requirements of the *Building Regulations 2000 Schedule 1* that he or she believes are unreasonable in the circumstances.

For a relaxation, the application should clearly set out the case and should say why it is believed that the requirement is too onerous and why it is unnecessary to comply with a particular aspect of it. This will enable the local authority to consider the relaxation request in the confidence that there is a sound rationale for approving it and will show the extent of the applicant's obligation to comply.

If a dispensation is being sought the applicant needs to justify why it is thought that the whole of a requirement is inappropriate or unreasonable, in the particular circumstances of the plans or building work. A dispensation effectively absolves the applicant from having to comply with a requirement and may be easier to justify than for a relaxation.

The power to relax or dispense is vested solely in the Local Authority even if an Approved Inspector has been appointed to carry out the operational building control function. Therefore, it makes good sense to discuss the proposal to seek a relaxation or dispensation with him or her before proceeding.

An application for a relaxation or dispensation can be made at any time (e.g. from sketch scheme stage to completion of the building works) and (unlike the Determination procedure) is not restricted to the full plans procedure provided by the Local Authority or the building control service provided by an Approved Inspector. Therefore, an application can be made where the Local Authority building notice procedure has been used.

However, the local authority cannot give a relaxation or dispensation if, before the application is made, they have given a notice under section 36 of the *Building Act 1984*, requiring the applicant to pull down, alter or remove work and this has not been complied with. By the same token, they cannot issue a section 36 notice if an application for relaxation or dispensation is still active.

At least 21 days before giving a relaxation or dispensation, the local authority must advertise the proposal in a local newspaper. The advertisement must indicate the situation and nature of the work, and must contain details of the requirement to be dispensed with or relaxed. It must also invite representations from the public on issues of public health and safety that might be compromised by the proposal, these to be given within 21 days from the date of publication. The cost of the advertisement is charged to the applicant.

There is no need to advertise the proposals where the work affects only the internal part of a building or where the proposal affects only premises adjoining the site. In the latter case, the need to publish a notice is replaced by the service of a notice containing similar details, on all adjoining owners or occupiers. Before making a decision the local authority must consider all representations given to it.

If the local authority refuses an application there is a right of appeal under the *Building Act 1984*, section 39, to the Secretary of State or the NAW against that decision, provided that it is lodged within 1 month of the date of notification of the refusal by the local authority. If the local authority do not notify the applicant of their decision within 2 months (or such longer period as agreed) an Appeal may be made to the Secretary of State or the NAW as if the local authority had refused, and in such a case the 1-month period for appeal runs from the end of the 2-month period.

When lodging an Appeal the following documents should be provided:

- the names and addresses of the parties involved, including any agents
- details of the Local Authority or Approved Inspector providing the building control service

- the full address of where the proposed building work will be or has been carried out
- a statement setting out details of the building, the building work and the matter in dispute
- a statement setting out the case for either relaxing or dispensing with the particular requirement of the Regulations
- a copy of the plans of the building work and any other documents which have been submitted to the Local Authority with the relaxation or dispensation application
- a copy of all relevant correspondence with the Local Authority involved, including the Local Authority's notice of refusal. If an Approved Inspector is undertaking the building control service, a copy of the correspondence with him or her should be included
- a copy of any listed building consent if required for the proposed work and any associated planning permission relevant to the listed building
- a copy of any other documents supporting the case for relaxation or dispensation, including calculations
- where appropriate, a location or block plan and photographs of the proposed work to illustrate particular points.

The appeal should be sent to the ODPM if the proposal is in England or the NAW if it is in Wales. No fees are payable for an appeal, and as there is no model application form, the application should be in the form of a letter.

Once the application is accepted it will be acknowledged, and will be copied to the Local Authority (and the Approved Inspector if one has been appointed to carry out the building control function) for its comments. The Local Authority (and Approved Inspector, if appropriate) will be given 21 days to respond and any responses will be copied to the applicant. The applicant has 21 days in which to respond to any comments, by writing to the Secretary of State or NAW as appropriate. Any response should be copied to the Local Authority (and the Approved Inspector if appropriate).

The Secretary of State or the NAW will give detailed consideration to the documentation submitted and will give the applicant and the building control body a decision. Notification of the decision will be by letter and this will be copied simultaneously to the Local Authority (and Approved Inspector if appropriate).

The Secretary of State or NAW does not make site visits. The appeal is considered on the basis of the written representations. If it is necessary to obtain further information from either party during the technical appraisal, the Secretary of State or NAW will seek this in writing and will copy the information received to the other party.

The Secretary of State or NAW aims to issue a decision within 3 months of the receipt of all documentation. Therefore, the whole process will take about 6 months on average with the most complex cases taking longer.

6.3.4 Determinations following conditional approvals by local authorities

Where a full plans application has been made to a local authority and it considers that the plans are defective because they show a contravention of the Regulations, it is empowered under section 16 of the *Building Act 1984* to either reject the plans, or pass them subject to certain conditions (see Section 5.2.2). A conditional approval can be granted on either or both of the following conditions:

- that specified modifications should be made to the deposited plans
- that further specified plans must be deposited.

This provides the local authority with a useful means of expediting an approval where the alternative is an outright rejection of the plans; however, any request by the applicant for the imposition of a condition must be in writing, as must the acceptance of any condition.

It is usually assumed that because the applicant's consent has been given to the imposition of conditions that this is unlikely to lead to a subsequent application for a Determination. However, it may be the case that although the applicant had misgivings about the imposition of particular conditions, it was thought to be prudent at the time to agree to the conditions in order to secure a valid full plans approval. The alternative to refusing to accept certain conditions would be for the plans to be rejected. Since the Secretary of State and the NAW usually consider applications for determinations after the local authority has come to a decision on a full plans application, they are prepared to consider determination applications relating to the imposition of conditions.

6.4 Approved Inspectors powers to control contravening works

Approved Inspectors are not empowered to enforce the Building Regulations (this can only be done by local authorities); however, there are powers in section 52 of the *Building Act 1984 and* Part VI of the *Building (Approved Inspectors etc) Regulations 2000*, which can result in the cancellation of the initial notice and reversion of the work to the local authority, resulting in the resurrection of their full enforcement powers.

Reversion of the work to the local authority for enforcement is a final option rarely, if ever, used in practice. It will only arise where the Approved Inspector has tried all other courses of action without success, and feels that the extent of the contravention is so serious that the threat of cancellation of the initial notice is the only option left.

In such circumstances, under section 52 of the *Building Act 1984* the Approved Inspector must be of the opinion that there is a contravention of the regulations and the procedure is activated by the service of a notice of contravention on the person carrying out the work. Under regulation 19 of the *Building (Approved Inspectors etc) Regulations* 2000, the notice must specify:

- those requirements of the Building Regulations which the Approved Inspector feels are not being complied with
- the location of the non-compliant work.

The notice will inform the client that unless the contravention is rectified within a period of 3 months from its date of service, the initial notice will be cancelled. This gives the client a period of grace in which to instruct the builder to take down and remove the contravening work or to carry out alterations to make it comply.

Where the contravening works have not been rectified within the 3-month period the Approved Inspector will cancel the initial notice. This is done by serving a notice of cancellation on the Local Authority and the client (see Form 6 of the *Building (Approved Inspectors etc) Regulations 2000*). A typical form is shown in Figure 6.1.

6.4.1 Effect of cancellation of the initial notice by the Approved Inspector

Once the initial notice has been cancelled by the Approved Inspector in the circumstances described above under section 52 of the *Building Act 1984*, the Local Authority's powers to enforce the Building Regulations will revive unless another initial notice is served on the Local Authority and accepted by it (see section 20(1)(c) of the *Building (Approved Inspectors etc) Regulations 2000* and section 53(7) of the *Building Act 1984*). The new initial notice must relate to the work already carried out under the original initial notice.

Where another initial notice is not given and accepted by the local authority, it becomes responsible for enforcing the Regulations and must be provided after giving reasonable notice with:

- plans of the building work so far carried out
- plans of the work which has still to be completed
- plans referred to in any plans certificate accepted for the work.

The plans provided must be sufficient to enable the local authority to determine whether or not the Regulations have been, or are likely to be, contravened by the proposals.

Additionally, the local authority may serve notice on the person carrying out the work to cut into, lay open or pull down work so that it may ascertain whether any work already carried out and not covered by a final certificate contravenes the Regulations. A fee, which is appropriate to that work, will be payable to the local authority.

FORM 6

Section 52(1) of the Building Act 1984 ('the Act')

The Building (Approved Inspectors etc) Regulations 2000 ('the 2000 Regulations')

NOTICE OF CANCELLATION BY APPROVED INSPECTOR

To: (Note 1)

....

....

....

1.   This notice relates to the following work: (Note 2)

Location: ....

....

....

Description of work: ....

....

....

....

Use of any building: ....

....

2.   An initial notice dated ....(Note 3) has been given and the above work was specified in it.

3.   I am the approved inspector in relation to that work.

4.   I hereby cancel the initial notice.

5.   I gave notice to the person carrying out the work in accordance with regulation 19 of the 2000 Regulations and he failed to remedy the contravention within the prescribed period. The contravention is (Note 4) ....

....

....

....

Signature ....

Date ....

NOTES
(1)   Insert the name and address of the person to whom the notice is given. It must be given to the local authority and the person carrying out or intending to carry out the work.
(2)   Location and description of the work including the use of any building to which the work relates.
(3)   Insert date.
(4)   Delete this statement if it does not apply. If it applies, specify the provision of building regulations (including the specific requirement) which is contravened.

Figure 6.1 Typical notice of cancellation of an initial notice by an Approved Inspector.

The benefit of having a plans certificate can now be seen, since according to regulation 16 of the *Building (Approved Inspectors etc) Regulations 2000* the local authority is not permitted to serve a section 36 notice or institute proceedings for a contravention of the Building Regulations in relation to any work described in the certificate, provided that it has been carried out in accordance with the plans to which the certificate relates. However, even though work was covered by a plans certificate, there would not appear to be anything in the Regulations or Building Act to prevent the local authority from requiring proof that already covered up work did in fact comply with the Regulations.

Satisfying the requirements of the Regulations

7.1 General introduction

Chapter 3 describes the structure and nature of the building control system in England and Wales by analysing the *Building Act 1984*, the *Building Regulations 2000* and the *Building (Approved Inspectors etc) Regulations 2000*. It is this 'structure' which determines:

* what can be required by the regulations
* how the requirements can be met.

Under the provisions in section 1 of the Building Act 1984, at present the Secretary of State may make regulations with respect to:

* the design and construction of buildings
* the provision of services, fittings and equipment in, or in connection with, buildings.

The regulations may be made only for the purposes of:

(i)   securing the health, safety, welfare and convenience of persons in or about buildings (this includes other people who may be affected by buildings or matters connected with buildings)
(ii) furthering the conservation of fuel and power
(iii) preventing waste, undue consumption, misuse or contamination of water.

It should be noted that it is the ability to make regulations about welfare and convenience in point (i) above that has enabled the Secretary of State to control access and facilities for disabled people in (and about) buildings. At present, controls over

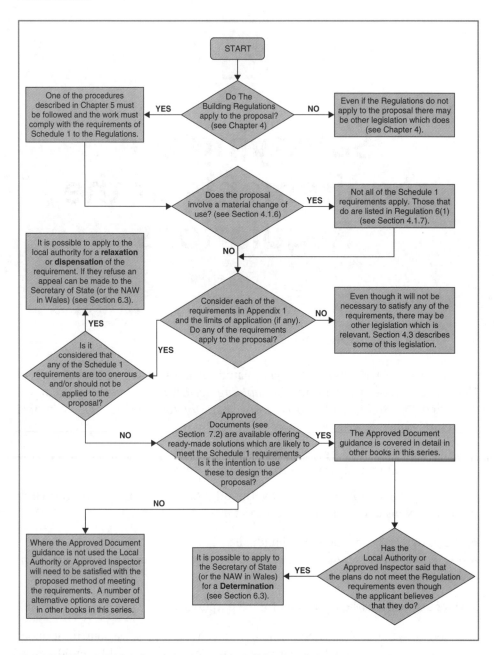

Flow chart 7.1 Satisfying the requirements of the Building Regulations.

water supply and water fittings are not contained in the Building Regulations but instead are set out in the *Water Supply (Water Fittings) Regulations 1999*. Thus they remain a separate set of requirements for developers to consider.

As is explained in Chapter 6, the Building Regulations impose mandatory requirements on people who propose to carry out building work with respect to the design and construction of buildings. They are phrased in functional terms with the intention of giving designers and builders flexibility in the way they comply therefore they contain no practical guidance regarding methods of compliance. Non-mandatory guidance is provided principally by Approved Documents and Harmonised Standards (British or European); however, there are other methods of demonstrating compliance, such as past experience of successful use, test evidence, calculations, compliance with European Technical Approvals, the use of CE-marked materials, etc.

Flow chart 7.1 gives an outline of the ways in which the regulations may be satisfied.

7.2 Approved Documents

7.2.1 Background

Under section 6 of the Building Act 1984 the Secretary of State may approve any document for the purpose of providing practical guidance with respect to the requirements of the regulations. Such documents (known as Approved Documents) are intended to be written and illustrated in comparatively straightforward technical terms and are updated at intervals after extensive consultation with a wide range of interested parties who are deemed to be representative of the particular issues covered by the Approved Document under preparation or revision. The actual work of identifying and considering the content of new and revised Regulations and Approved Documents is carried out by a body (the Building Regulations Advisory Committee, BRAC) appointed by the Secretary of State for the Office of the Deputy Prime Minister (ODPM) under powers contained in section 14 of the Building Act 1984. Members of BRAC are unpaid (although they are permitted to receive expenses incurred in connection with their work). Approved Documents can be produced directly by the ODPM or alternatively, the Secretary of State (after consultation) may approve guidance documents produced by other specialist bodies. To date there have been 16 Approved Documents issued by the ODPM. These cover all parts of Schedule 1 (see Appendix 1) and include an Approved Document which gives guidance on meeting the requirements of regulation 7 (*Materials and workmanship*). The latest Approved Document covering *Part P (Electrical Safety)* came into force on 1 January 2005.

The Secretary of State has also approved two other specialist Approved Documents entitled respectively:

- *Timber Intermediate Floors for Dwellings*, published by the Timber Research and Development Association
- *Basements for Dwellings*, published by the British Cement Association.

7.2.2 Critique of the current Approved Documents

Unfortunately, the Building Regulations and their associated Approved Documents have been getting more complex with every update, often requiring the services of specialist professionals (services engineers, fire engineers, etc.) to make sense of the provisions.

New areas of control are being introduced each year (a proposed Part Q covering electronic communications services is currently under discussion) and the scope of the existing regulations is being extended with each revision.

Furthermore, the current Approved Documents are principally of use in the design of extremely simple and straightforward buildings using mainly traditional techniques. For larger and more complex buildings it is usually better (and more efficient in terms of building design) to use other sources of guidance (British and European Standards, Building Research Establishment Reports, etc.) and although these other source documents are referenced in the Approved Documents no details of their contents or advantages of use are given. As an example of this, the current edition of Approved Document L2 (*Conservation of fuel and power – buildings other than dwellings*) contains numerous footnote references on virtually every page to no fewer than 56 other sources of guidance.

Additionally, it is often the case that the current Approved Documents fail to provide sufficient guidance just when it is needed; that is, when it is proposed to deviate from the simple solutions or attempt to design something slightly unusual. Arguably, this has the tendency to encourage adherence to traditional and (perhaps) unimaginative designs and details, and can actively discourage innovation in many modern building designs.

This series of books has been specifically developed to improve on the current situation by addressing different parts of the Building Regulations in separate volumes, thus enabling each part to be explored in detail.

Each volume expands on the Approved Documents by not only describing the traditional approach but also by making extensive reference to other sources of guidance contained in them. These 'alternative approaches' (as they are described in the Approved Documents) are analysed and the most critical parts of them are presented in the text with indications of where they can be used to advantage (over the traditional approach).

7.3 Using the Approved Documents

There is no legal obligation to use the Approved Documents when carrying out a building project to which the regulations apply. The applicant is perfectly entitled to choose whether or not to use all or some of the relevant Approved Documents or indeed, some parts of them. However, care should be taken in certain circumstances not to mix up Approved Document guidance with that from other guidance sources referred to in the specific Approved Document. For example, guidance on means of escape contained in Approved Document B (requirement B1) should not be 'mix and matched' with guidance contained in the relevant part of BS 5588 (*Fire precautions in the design, construction and use of buildings*). If the decision is made to use BS 5588, then the parts which are relevant to the means of escape should be followed rather than a mixture of the specific publication and the relevant sections of Approved Document B1.

Therefore, the legal obligation is to meet the requirements of the Regulations, not the Approved Document guidance. This leaves the designer or developer with the option of devising his or her own solution or of following other authoritative guidance. There may even be circumstances in which it would be unnecessary or inappropriate to follow the guidance in the Approved Document in full, or the Approved Document may offer no guidance on a particular situation being faced.

7.3.1 The legal status of the Approved Documents

If adherence to the Approved Document guidance is not mandatory, to what extent do they provide the designer with the confidence that he or she is indeed satisfying the requirements of the Regulations and what, if any, is the nature of their legal status? Section 7 of the Building Act 1984 provides that if the guidance in the Approved Documents is followed and it is alleged in any proceedings that the Regulations have been contravened – *'proof of compliance with such a document may be relied on as tending to negative liability'*. Conversely, if the guidance in the Approved Documents has not been followed and it is alleged that the Regulations have been contravened; that is, *'a failure to comply with a document that at that time was approved for the purposes of that provision may be relied upon as tending to establish liability'*. Put more simply, the onus will be on the designer or developer to establish that he or she has met the requirements in some other way and he or she may be called upon by the building control body to demonstrate this by other means. In the case of *Rickards v. Kerrier District Council* (1987) it was held that if the local authority proved that the works did not comply with the Approved Document then it was for the appellant to show compliance with the regulations.

It should be noted that proof of non-compliance with an Approved Document does not necessarily mean that the regulation has been contravened.

7.3.2 Extent of compliance

Therefore, if the decision is taken not to use the Approved Document guidance it is necessary to be clear as to the extent of the compliance that is demanded by the regulations. Building Regulations do not guarantee that a building will be fit for its purpose or that building work will be good value for money and done to a high standard, etc. Regulation 8 defines the limitations of the regulations as follows:

> *Parts A to D, F to K and N (except for paragraphs H2 and J6) of Schedule 1 shall not require anything to be done except for the purpose of securing reasonable standards of health and safety for persons in or about buildings (and any others who may be affected by buildings, or matters connected with buildings).*

From a reading of regulation 8 it is evident that for most parts of Schedule 1, compliance with the regulations means securing reasonable standards of health and safety in the building in question, as a result of the controlled building work. These health and safety limitations are not relevant to the following parts of Schedule 1, but instead are replaced by the other considerations listed below. All these matters are amongst the purposes, other than health and safety that may be addressed by Building Regulations under the *Building Act 1984, section 1(1)*:

- Part E (*Resistance to the passage of sound*). Here, the issue of welfare and convenience must be addressed since it is clearly unpleasant and inconvenient to have unwanted sound entering a building through the intervening walls and floors where the occupant is unable to control the level of sound emanating from a neighbour. Interestingly, this concept was extended to sound levels between certain rooms in the same building by virtue of the *Building (Amendment) (No. 2) Regulations 2002*.
- Part L (*Conservation of fuel and power*). This is specifically mentioned in the *Building Act 1984, section 1(1)(b)* as being amongst the purposes for which the Secretary of State may make building regulations. In their original form, the regulations were related strictly to measures which could be incorporated into buildings to reduce the consumption of energy. Since the Kyoto agreement the emphasis has been not only on energy efficiency but also on measures to reduce carbon dioxide emissions.
- Part M (*Access to and use of buildings*). Until the latest amendment (which came into force on 1 May 2004) Part M had always been overtly related to access and facilities for *disabled* people. The 2003 changes to Part M resulted in a new title and a different emphasis, all references to disabled people having been dropped in the requirements, since making buildings accessible and usable means that they are suitable (and convenient) for all people. Hence, welfare and convenience are the key issues when considering accessibility.
- Requirement H2 (*Wastewater treatment systems and cesspools*). A close examination of requirement H2 reveals that it deals with a range of issues including:
  (a) siting and construction (not to be prejudicial to health)
  (b) prevention of contamination of watercourses and supplies
  (c) access for emptying and maintenance
  (d) the ability to function to a sufficient standard for the protection of health in the event of a power failure (relevant to motorised systems)

(e) the provision of information concerning maintenance.

Clearly, there are health and safety issues within H2 as well as others concerning prevention of pollution, access and provision of information. Therefore, the exclusion of H2 from the need to comply with health and safety issues by virtue of regulation 8 does not sit comfortably with the wording of H2 and this would appear to be a case of one regulation contradicting another.

- Requirement J6 (*Protection against pollution*). For J6, compliance is limited to measures necessary to minimise the risk of oil escaping and causing pollution in oil storage tanks and connecting pipework. It covers the construction and protection of oil storage tanks and connecting pipework, and the need to provide permanent notices containing information on how to respond to an oil escape.

7.4 Methods of compliance

7.4.1 Level of performance

Most Approved Documents give guidance in a number of different ways. When using them, it is most important to achieve a level of performance that will be seen to satisfy the relevant regulation requirement. The level of performance necessary is usually stated at the beginning of the Guidance section of the Approved Document (i.e. after the reiteration of the regulation requirement). For example, in the current edition of Approved Document M the performance level is stated as:

Performance
In the Secretary of State's view the requirements of Part M will be met by making reasonable provision to ensure that buildings are accessible and usable.
People, regardless of disability, age or gender, should be able to:

a. *gain access to buildings and to gain access within buildings and use their facilities, both as visitors and as people who live or work in them;*
b. *use sanitary conveniences in the principal storey of a new dwelling.*

The provisions are expected to enable occupants with disabilities to cope better with reducing mobility and to 'stay put' longer in their own homes.
The provisions are not necessarily expected to facilitate fully independent living for all people with disabilities.

This statement of performance is vital to an understanding of the extent to which compliance is required. For example, it makes it clear that it applies to all kinds of people (*regardless of disability, age or gender*), that it applies to people who work in a building as well as to visitors to that building, and that it does not mean that all dwellings have to be constructed specifically for people with disabilities.

Although a stated level of performance enhances the regulation requirement by giving additional information to designers, it does not supply ready-made solutions for any particular design scenario. For this the Approved Documents supply information in the form of *Technical Solutions*, and where these are not acceptable *Alternative Approaches* are also often given as described below.

7.4.2 Technical solutions

The most common and (to many people), the most useful way of providing guidance is when the Approved Documents describe (often with illustrations) particular methods of construction. These so-called *technical solutions* give details of some of the more widely used forms of construction which achieve an acceptable level of performance. Although they generally give detailed guidance, in some cases they are written to give sufficient flexibility to adapt them to suit a method of construction which the designer prefers.

This approach is well illustrated in Approved Document L1 (*Conservation of fuel and power – dwellings*). In this document three methods are shown for demonstrating reasonable provision for limiting heat loss through the external envelope (including the ground floor) of the dwelling:

(a) an Elemental Method
(b) a Target U-value Method
(c) a Carbon Index Method.

The *Elemental Method* allows the designer, for example, to select from a range of commonly used forms of construction for elements, such as floors, walls and roofs, by utilising a series of tables giving different insulation thicknesses relevant to a variety of insulation materials. This minimises the need to carry out lengthy calculations but restricts the methods of construction available to those illustrated in the tables. Other restrictions will apply to the design if this approach is adopted. For example, heating the dwelling by electricity will not be permitted, and window and door areas will be limited.

On the other hand, using the *Target U-value Method* allows greater flexibility than the Elemental Method (within certain limits) in selecting the areas of windows, doors and rooflights, and the insulation levels of individual elements in the building envelope. It also takes into account the efficiency of the heating system and enables solar gain to be addressed. Additionally, it can be used for any heating system. It requires that some simple calculations be carried out.

The *Carbon Index Method* is aimed at providing even more flexibility in the design of new dwellings whilst achieving similar overall performance to that obtained by following the Elemental Method. The method involves fairly complicated calculations which attempt to model the annual production of carbon dioxide produced by the dwelling adjusted for floor area, thus making the carbon index figure independent of dwelling size for a given built form. In theory, this should allow any form of construction to be considered provided, of course, that certain risks (such as surface and interstitial condensation) can be addressed.

From this example it can be seen that the Approved Document can cater for many different levels of expertise from simply selecting forms of construction and material thicknesses from standard tables to using formulae, which attempt to model the carbon dioxide production of the dwelling.

Sometimes the technical solutions contained in an Approved Document are entirely related (and limited) to a particular form or mode of construction. For example, the 2004 edition of Approved Document A (*Structure*) includes guidance on the design of the structural elements of residential buildings of traditional masonry construction. The Document recognises the fact that there are other suitable forms of construction in use in the housing sector (such as timber framed housing) some of which have been in common use for many years and have demonstrated an adequate performance in compliance with the requirements of Paragraph A1 of Schedule 1 to the regulations. Such alternative forms include prefabricated timber, light steel and pre-cast concrete framed construction. A number of guidance documents relating to these alternative forms are presently being developed by industry and it is the intention to reference these in Approved Document A as soon as they become available and are approved by the Secretary of State. In the meantime, any designer adopting a non-traditional approach to housing will have to prove, by other means, that the design complies with regulation A1.

7.4.3 Alternative approaches: the use of named standards and technical approvals

What can be done if a designer decides that a technical solution given in an Approved Document is inappropriate (too restrictive or inflexible; not allowing the use of innovative materials or construction methods; does not take into account the intricacies of the building design; etc.)? In many cases the Approved Documents will give details of Alternative Approaches that can be adopted. These are usually based on the relevant recommendations of certain technical specifications, such as named standards, technical approvals and/or other 'official' guidance documents. Adoption of these alternatives may also give the designer an opportunity to use a more complex procedure to 'fine tune' the design solution. For example, underground surface water drainage systems may be designed by following the guidance given in section 3 of H3 in the 2002 edition of Approved Document H (paragraphs 3.1 to 3.35). Paragraph 3.36 gives the alternative approach as follows:

> The requirement [*i.e. for discharge of the surface water*] can also be met by following the relevant recommendations of *BS EN 752 – 4 Drain and sewer systems outside buildings*. The relevant clauses are in Part 4 *Hydraulic design and environmental considerations* clauses 3 to 12 and national annexes NA, NB and ND to NI. BS EN 752, together with BS EN 1295 and BS EN 1610 contains additional detailed information about design and construction.

It will be seen in this example that only certain clauses are relevant to building regulation compliance. This is because building regulations are made for specific purposes; that is, health and safety, energy conservation, and the welfare and convenience of disabled people. Alternative approaches using named standards (such as those listed in the example) may provide relevant guidance to the extent that they relate to these considerations. However, they may also address other aspects of performance such as serviceability, or aspects which although they relate to health and safety are not covered by the Regulations.

Where reference is made to a named standard or technical approval in an Approved Document the relevant version of the standard will be the one listed at the end of the publication. However, if this version has been revised or updated by the issuing standards body, the new version may be used as a source of guidance provided it continues to address the relevant requirements of the Regulations.

7.5 Materials and workmanship

So far, the methods of compliance described relate principally to the design of the building. It should be noted that the way in which the building work is carried out and the adequacy of the materials used are also addressed by the Building Regulations.

Regulation 7 (*Materials and workmanship*) states that:

Building work shall be carried out –

(a) with adequate and proper materials which –
  (i) are appropriate for the circumstances in which they are used,
  (ii) are adequately mixed and prepared, and
  (iii) are applied, used or fixed so as adequately to perform the functions for which they are designed; and
(b) in a workmanlike manner.

7.5.1 Fitness of materials

The Approved Document to support regulation 7 gives the following examples of how the requirements of regulation 7 may be satisfied with regard to the fitness of materials:

- By conforming with the relevant provisions of an appropriate British Standard (BS) or to the national technical specifications of other Member States which are contracting parties to the European Economic Area. (In this latter case, the onus is on the person intending to use the material to show that it is equivalent to the requirements of the relevant BS, if necessary by providing a translation.)

- By being covered by a national or European certificate issued by a European Technical Approvals issuing body (such as a material covered by a British Board of Agrément Certificate). The certificate will indicate which provisions of the regulations are so covered.
- By bearing a CE marking confirming that the material conforms with a harmonised European Standard or European Technical Approval together with the appropriate attestation procedure. A CE-marked product can only be rejected by a building control body if:
  (a) its performance is not in accordance with its technical specification or
  (b) where a particular declared value or class of performance is stated for a product, the resultant value does not meet Building Regulation requirements.
  In this case the burden of proof is on the controlling body, which is obliged to notify the Trading Standards Officer so that the UK Government can notify the Commission.
- By conforming to the requirements of an independent certification scheme, such as the kitemark scheme operated by the British Standards Institution (BSI), provided that the certification scheme has been suitably accredited (e.g. by UK Accreditation Service, UKAS).
- By the use of tests, calculations or other means which show that the material is capable of performing its function. The tests must be carried out in accordance with the recognised criteria such as those carried out by testing laboratories accredited by UKAS.
- By demonstrating that past experience of a material or method of use in a building has shown that the material is capable of adequately performing its function.

7.5.2 Adequacy of workmanship

Adequacy of workmanship, like that of materials, may be established in a number of ways including conformity with a national standard or technical specification. In this context, BS 8000: *Workmanship on Building Sites* may be useful since it gathers together guidance from a number of other BSI Codes and Standards.

Workmanship which is covered by a scheme complying with BS EN ISO 9000: *Quality management and quality assurance standards* will demonstrate an acceptable standard since these schemes relate to processes and products for which there may also be a suitable British or other technical standard. A number of such schemes have been accredited by UKAS. There are also a number of independent schemes for accreditation and registration of installers of materials, products and services, and these ensure that work has been carried out to appropriate standards by knowledgeable contractors. For example, an unvented hot water storage system may be installed by a person who holds a current Registered Operative Identity Card issued by:

- the Construction Industry Training Board (CITB)
- the Institute of Plumbing

- the Association of Installers of Unvented Hot Water Systems (Scotland and Northern Ireland) or
- other equivalent bodies.

In some cases, where it can be demonstrated that a method of workmanship has been successful in the past in satisfying the requirements of the regulations, this past experience may be relied upon to ensure that the method is capable of producing the intended standard of performance.

7.6 Approved Documents and alternative solutions

7.6.1 Introduction

As has been explained above, Approved Documents represent a convenient way of providing relatively simple solutions to a wide range of building design problems. However, it is often the case that they fail to provide sufficient guidance just when it is needed; that is, when it is proposed to deviate from the simple solutions or attempt to design something slightly unusual. Arguably, this has the tendency to encourage adherence to traditional and (perhaps) unimaginative designs and details, and can actively discourage innovation in many modern building designs. The very nature of the *Technical Solutions* (see paragraph 7.4.2) may make it difficult for a designer to come up with alternative design solutions, which whilst satisfying the functional requirements of the Regulations do not follow the Approved Document recommendations in full.

7.6.2 Access Statements

One way of justifying the use of alternative solutions to the Approved Document guidance can be illustrated by considering the use of Access Statements. Where relevant, a project must comply with the provisions of Part M *Access to and Use of Buildings* and guidance on compliance can be found in the 2004 edition of Approved Document M. This guidance may appear to be more prescriptive than that contained in other Approved Documents because much of it is based on BS 8300:2001 *Design of buildings and their approaches to meet the needs of disabled people – Code of Practice* and is by derivation based largely on the ergonomic studies carried out to support the BS. In the case of extensions and material changes of use of buildings other than dwellings, and particularly in the case of historic buildings it may be extremely difficult to satisfy the recommendations of Approved Document M due to the prescriptive nature of the guidance. Access Statements can be used at both the planning and building control stages of a project in order to clarify the accessibility issues on which the design is based. For example, in the case of an extension or material change of use to a multi-occupancy building it may be very difficult to provide a fully accessible access route where the landlord will not permit alterations to the common parts. Access Statements can be useful in such circumstances and can allow an applicant to identify the constraints imposed by the existing structure and its immediate environment, and to propose compensatory measures where full access proves to be impracticable or unreasonable.

The following examples show how an Access Statement can be used in varying circumstances to provide additional information to the building control body to justify situations where it may not be possible to follow the Approved Document recommendations in full:

(1) A Statement can be used by an applicant to identify buildings or particular parts of buildings where it would be either reasonable for access to be restricted or unreasonable to expect certain groups of people to require access. For example, where hazardous materials are handled, or in certain manufacturing processes, or in areas where archiving and bulk-handling processes are carried out, which might create hazards for children, some disabled people or frail elderly people.

(2) In the case of a relevant material change of use, it may be impracticable to make the existing principal entrance or any other appropriate existing entrance suitable for use by particular groups of people, or to provide a new entrance which is suitable. The Access Statement can give the reasons why it is not practicable to adjust the existing entrance or provide a suitable new entrance.

(3) In the case of an extension where it is not intended to provide a fully compliant-independent access, and there are no accessible toilets in another part of the same building, the Access Statement can be used to indicate why a fully compliant-independent access is not considered reasonably practicable.

7.6.3 Planning applications and building control submissions

Where planning permission is required for a project, the ODPM publication *Planning and Access for Disabled People – A Good Practice Guide* recommends provision of an Access Statement to identify:

• the philosophy and approach adopted in access design
• the key issues of the particular scheme
• the sources of advice and guidance used.

An additional benefit of providing an Access Statement is that it should set out at the time of the planning application most of the information needed by a building control body, thus assisting the dialogue between the applicant and building control. Therefore, the Access Statement provided for planning purposes should be seen as a basis for (and a complement to) an Access Statement provided for building control purposes, rather than as a separate document. It can then be provided at the time plans are deposited, a building notice is given or details of a project are given to an Approved Inspector. It may also be beneficial to maintain and update the Access Statement as the building work progresses in order to provide the eventual user of the building, who may have ongoing obligations under the Disability Discrimination Act, with a record of decisions made which had an impact on accessibility, and of the reasons for such decisions.

7.6.4 Access Statements and Approved Document guidance

At its very simplest, an Access Statement might record that the intention of the client, designer or design team ('the applicant') was to comply where appropriate with the guidance in Approved Document M, and to indicate in what respects it was considered appropriate. Where an applicant wishes to depart from the guidance in Approved Document M, either to achieve a better solution using new technologies (e.g. infrared-activated controls), to provide a more convenient solution, or to address the constraints of an existing building, the Statement should set out the reasons for departing from the guidance and the rationale for the design approach adopted. Examples of evidence that might be cited to support such an approach might include:

- application of the recommendations in BS 8300 where these differ from the provisions, or are not covered, in Approved Document M
- results of current validated research (perhaps published in the last 5 years)
- outcome of consultations with other parties, such as Conservation Officers, English Heritage or the Welsh Historic Monuments Executive Agency (CADW), Local Access Officers, etc.
- convincing arguments that an alternative solution will achieve the same, a better, or a more convenient outcome.

7.6.5 Conclusions

Although this approach is aimed specifically at access to and use of buildings, there is no reason why the technique cannot be applied in many other cases where the decision is taken to avoid use of the Approved Document guidance.

The key examples of evidence listed in paragraph 7.6.4 (i.e. reference to other guidance sources, results of recent research, consultation with expert bodies, and arguments to prove that the proposed solution is at least as good as the Approved Document guidance) can apply, in principle, to any of the Approved Documents and have been used many times in the past with excellent results. The designer should never feel that he or she has to use Approved Documents; however, they must be prepared to put forward a sound case to the building control body if the designer's alternative solution is to be accepted.

PART 4

Appendices

Appendix 1

Commentary on the Building Regulations 2000 (as amended)

Introduction

In the following commentary, the requirements of the *Building Regulations 2000 (SI 2000/2531)* have been combined with subsequent amendments (*SI 2001/3335; SI 2002/440, SI 2002/2871, SI 2003/2692, SI 2004/1465* and *SI 2004/1808*) to produce a consolidated version. This will provide the reader with a complete updated text and will avoid the need for constant cross-referencing amongst the documents listed above.

The notes in this commentary should be read in conjunction with the main text.

Building Regulations 2000	Commentary
Building Regulations 2000 (SI 2000/2531) *Made*: 13 September 2000 *Laid before Parliament*: 22 September 2000 *Coming into force*: 1 January 2001 The Secretary of State, in exercise of the powers conferred upon him by sections 1(1), 3(1), 8(2), 35 and 126 of, and paragraphs 1, 2, 4, 7 and 8, 10 of Schedule 1 to, the Building Act 1984 and of all other powers enabling him in that behalf, after consulting the Building Regulations Advisory Committee and such other bodies as appear to him to be representative of the interests concerned in accordance with section 14(3) of that Act, hereby makes the following Regulations:	The 2000 Regulations replaced the *Building Regulations 1991 (SI 1991/2768)* as amended. However, the 1991 Regulations will continue to apply in relation to work which was notified to a local authority either directly (via a building notice or as a full plans application) or indirectly (via an Approved Inspector's initial notice or amendment notice) before the 1 January 2001. This does not prevent the plans (or initial notice) from lapsing if work has not commenced within 3 years of deposit. Any subsequent application would then be subject to the provisions of the new Regulations. The various amending Regulations have also been subject to certain transitional arrangements.

Building Regulations 2000	Commentary
PART I: GENERAL **1 Citation and commencement** (1) These Regulations may be cited as the Building Regulations 2000 and shall come into force on 1 January 2001.	
2 Interpretation	The interpretation section should be read in conjunction with the *Building Act 1984, section 126* (General interpretation) and the *Interpretation Act 1978* (which applies generally to all Acts of Parliament unless a contrary intention is expressed in another Act). The definitions contained in this section apply throughout the regulations unless specifically excluded elsewhere.
(1) In these Regulations unless the context otherwise requires: 'the Act' means the Building Act 1984;	The *Building Act 1984* contains the power to make building regulations and many administrative provisions, which govern the running of the building control system in England and Wales. The amending Regulations listed above also contained amendments to the *Building Act 1984* and these are referred to in the commentary text below. All references to 'the Act' in the text below are to the *Building Act 1984*.
'amendment notice' means a notice given under section 51A of the Act;	An 'amendment notice' is given by an Approved Inspector and his client when it is proposed to vary controlled building work that was the subject of an initial notice (i.e. work controlled by an Approved Inspector, not the local authority) and where the scope of the works has changed to such an extent that the original initial notice no longer truly reflects the work actually being carried out. Section 51A provides the rules for serving the amendment notice on the local authority.

Building Regulations 2000	**Commentary**
'building' means any permanent or temporary building but not any other kind of structure or erection, and a reference to a building includes a reference to part of a building;	The definition of 'building' in the regulations is narrower than that contained in the *Building Act 1984*, the purpose being to restrict the scope of the Regulations to what are commonly thought of and referred to as buildings. The Building Act definition is necessarily couched in very wide terms so as to provide the local authority with the necessary powers to deal with, for example, dangerous structures and demolitions. In the past, doubt has been cast over the status of such structures as residential park homes and marquees. Provided that a residential park home conforms to the definition given in the *Caravan Sites and Control of Development Act 1960* (as augmented by the *Caravan Sites Act 1968*) it is exempt from the definition of 'building' contained in the regulations, and according to the *Manual to the Building Regulations* a marquee is not regarded as a building.
'building notice' means a notice given in accordance with regulations 12(2)(a) and 13;	An alternative to giving the local authority 'full plans' for determining compliance with the regulations. A building notice cannot be used for work to a building which is to be put to a 'relevant use' (see note to regulation 12 below).
'building work' has the meaning given in regulation 3(1);	The regulations apply only to 'building work' as defined in regulation 3(1). Even so, certain buildings are exempt from the requirements of the regulations (such as those occupied by the Crown, and those belonging to Statutory Undertakers, the UK Atomic Energy Authority and the Civil Aviation Authority) so the definition does not apply to such buildings.
'controlled service or fitting' means a service or fitting in relation to which Part G, H, J, L or P of Schedule 1 imposes a requirement;	This means the installation of bathroom fittings, hot water storage systems, sanitary conveniences, drainage and waste disposal systems, combustion appliances; and replacement doors, windows and rooflights, space heating and hot water boilers and hot water vessels, and work to certain electrical.installations.

Building Regulations 2000	Commentary
'day' means any period of 24 hours commencing at midnight and excludes any Saturday, Sunday, Bank holiday or public holiday;	The definition of 'day' is needed with respect to the periods of notice required by virtue of regulation 15.
'dwelling' includes a dwelling-house and a flat;	Although this does not actually define what a dwelling is, decided case law would indicate that the dwelling does not have to be occupied to be so classified. Additionally, it has been held that cooking facilities do not have to be available for the premises to come within the definition of 'dwelling'.
'dwelling-house' does not include a flat or a building containing a flat;	This definition clearly delineates houses from flats and allows separate provisions to be made for each.
'electrical installation' means fixed electrical cables or fixed electrical equipment located on the consumer's side of the electricity supply meter;	This definition is needed in relation to Part P, Electrical Safety, of Schedule 1.
'energy rating' of a dwelling means a numerical indication of the overall energy efficiency of that dwelling obtained by the application of a procedure approved by the Secretary of State under regulation 16(2) of these Regulations;	Regulation 16(2) requires that when a person carries out building work which involves the creation of a new dwelling (dwelling-house, flat or maisonette) he shall give notice of the energy rating of the dwelling to the local authority. The energy rating must be calculated in accordance with a standard procedure approved by the Secretary of State and described in Approved Document L1.
'European Technical Approval issuing body' means a body authorised by a member state of the European Economic Area to issue European Technical Approvals (a favourable technical assessment of the fitness for use of a construction product for the purposes of the Construction Products Directive);	An example of the use of this definition may be found in regulation 13(3) in relation to building work which involves the installation of insulating material into the cavity walls of a building.
'extra-low voltage' means voltage not exceeding: (a) in relation to alternating current, 50 volts between conductors and earth; or on relation to direct current, 120 volts between conductors;	This definition is needed in relation to the limits on application for Part P, Electrical safety, of Schedule 1.

Building Regulations 2000	Commentary
'final certificate' means a certificate given under section 51 of the Act;	A certificate given by an Approved Inspector under section 51 of the Building Act 1984 to indicate that a project has been successfully completed in accordance with the regulations. Acceptance of this certificate by the local authority cancels its powers to take action against a person for contravention of the regulations.
'flat' means separate and self-contained premises constructed or adapted for use for residential purposes and forming part of a building from some other part of which it is divided horizontally;	This definition covers maisonettes as well as flats. The main problem with this definition lies in the interpretation of the phrase *'separate and self-contained premises'*. In the past questions have arisen over the status of staff flats (e.g. over licensed premises) where these are used exclusively by the pub manager or his staff, particularly in relation to the application of the resistance to the passage of sound requirements of Part E. Where such 'flats' are used for residential purposes and they are separately approached and have separate lockable front doors it is reasonable to assume that Part E does apply and that the premises described are truly flats within the definition. Decided case law is relatively quiet on the subject of *'separate and self-contained premises'*, however in the case of *Malekshed v Howarde Walden Estates Ltd, Times Law Reports, 9 July 2001*, it was held that a substantial terraced house in Central London physically linked by a basement to a mews property at the rear could reasonably be called a house, although in this case both properties had been demised together under the terms of a Head Lease.
'floor area' means the aggregate area of every floor in a building or extension, calculated by reference to the finished internal faces of the walls enclosing the area, or if at any point there is no such wall, by reference to the outermost edge of the floor;	This definition is principally of use in Schedule 2 – Exempt buildings and work.

Building Regulations 2000	Commentary
'fronting' has the meaning given in section 203(3) of the Highways Act 1980;	This term appears in regulation 12 with the effect of requiring the deposit of full plans (where the local authority is used for building control) for the erection of a building fronting onto a private street. 'Fronting' can also mean being adjacent to.
'full plans' means plans deposited with a local authority for the purposes of section 16 of the Act in accordance with regulations 12(2)(b) and 14;	The deposit of full plans is an alternative to the giving of a building notice when using the local authority route for building control. Full plans *must* be deposited for certain kinds of work (see note to regulation 12 below). The *Building Act 1984, section 126* defines 'plans' as including drawings of any description and specifications or other information in any form. In certain circumstances and for certain kinds of building work it is unnecessary to deposit full plans or give a building notice (see note to regulation 12 below).
'height' means the height of the building measured from the mean level of the ground adjoining the outside of the external walls of the building to the level of half the vertical height of the roof of the building, or to the top of the walls or of the parapet, if any, whichever is the higher;	This definition is principally of use in Schedule 2 – Exempt buildings and work.
'independent access' means, in relation to a part of a building (including any extension to that building), a route of access to that which does not require the user to pass through any other part of the building.	See also Paragraph M2 of Part M, Access and use in Schedule 1 for an example of where this definition applies.
'initial notice' means a notice given under section 47 of the Act;	The deposit of an initial notice under *section 47* of the *Building Act 1984* by an Approved Inspector, and its acceptance by the local authority, effectively removes control of the work from the local authority and places it under the control of the Approved Inspector. *Section 47* describes the conditions under which the initial notice may be given and received.

Building Regulations 2000	Commentary
'institution' means an institution (whether described as a hospital, home, school or other similar establishment) which is used as living accommodation for, or for the treatment, care or maintenance of persons: (a) suffering from disabilities due to illness or old age or other physical or mental incapacity, or (b) under the age of 5 years, where such persons sleep on the premises;	Specific reference to the term 'institution' may be found in *regulation 5(b) Meaning of material change of use*. The most important consideration is that the occupants of such buildings must sleep on the premises and so day care centres, etc., are not included. It should be noted that health service buildings are now subject to the full substantive and procedural provisions of the Building Regulations (as they are no longer regarded as Crown buildings).
'low voltage' means voltage not exceeding: (a) in relation to alternating current, 1000 volts between conductors or 600 volts between conductors and earth; or (b) in relation to direct current, 1500 volts between conductors or 900 volts between conductors and earth;	This definition is needed in relation to the limits on application for Part P, Electrical safety, of Schedule 1.
'material alteration' has the meaning given in regulation 3(2);	Building work which constitutes of a 'material alteration' to a building is subject to control under the regulations. The definition in regulation 3(2) is phrased in general terms and can cause problems of interpretation. The *Manual to the Building Regulations* makes it clear that works of repair (replacement, redecoration, routine maintenance and making good) do not come under the definition.
'material change of use' has the meaning given in regulation 5;	Nine cases of material change of use are given in regulation 5 and these differ substantially from the definition given under planning law in the *Town and Country Planning Act 1990, s 55*, so approval under the Building Regulations may be required for a change of use where planning approval is not required and vice versa.
'private street' has the meaning given in section 203(2) of the Highways Act 1980;	
'public body's final certificate' means a certificate given under paragraph 3 of Schedule 4 to the Act;	See also Form 12 in Schedule 2 of the Building (Approved Inspectors etc) Regulations 2000.

Building Regulations 2000	Commentary
'public body's notice' means a notice given under section 54 of the Act;	The effect of *section 54 of the Building Act 1984* is to enable designated bodies to self-certify their own work (in a similar manner to the system operated by Approved Inspectors). Although in essence, this exempts designated public bodies from the procedural requirements of the regulations, building work carried out by them must still comply with the substantive provisions. To date only the Metropolitan Police Authority has been designated as a public body (see regulation 10 below).
'room for residential purposes' means a room or suite of rooms, which is not a dwelling-house or flat and which is used by one or more persons to live and sleep and includes a room in a hostel, an hotel, a boarding house, a hall of residence or a residential home, whether or not the room is separated from or arranged in a cluster group with other rooms, but does not include a room in a hospital, or other similar establishment, used for patient accommodation and, for the purposes of this definition, a 'cluster' is a group of rooms for residential purposes which is: (a) separated from the rest of the building in which it is situated by a door which is designed to be locked; (b) not designed to be occupied by a single household;	It should be noted that the examples of accommodation encompassed by the expression *'room for residential purposes'* do not constitute an exhaustive list, so rooms or suites of rooms in some other sorts of building may also be covered by the expression.
'shop' includes premises: (a) used for the sale to members of the public of food or drink for consumption on or off the premises, (b) used for retail sales by auction to members of the public, (c) used by members of the public as a barber or hairdresser, or for the hiring of any item, and (d) where members of the public may take goods for repair or other treatment.	The definition of shop for the purposes of the regulations is quite wide and includes some uses which would not normally be so regarded such as pubs, restaurants and auction houses. Again, the list is not exhaustive (as is denoted by the use of the word 'includes' in the first line, so other types of shops where goods of various kinds are sold to the public would also be covered by the expression.

Building Regulations 2000	Commentary
(2) In these Regulations 'public building' means a building consisting of or containing: (a) a theatre, public library, hall or other place of public resort; (b) a school or other educational establishment not exempted from the operation of building regulations by virtue of section 4(1)(a) of the Act; or (c) a place of public worship; but a building is not to be treated as a place of public resort because it is, or it contains, a shop, storehouse or warehouse, or is a dwelling to which members of the public are occasionally admitted.	Specific reference to the term 'public building' may be found in *regulation 5(e) Meaning of material change of use*. Additionally, private theatre and cinema clubs and bingo halls (covered by the phrase *'other places of public resort'* in 2(2)(a)) for members only, would not normally be regarded as 'public buildings'. This has given rise to difficulties in the past where admission procedures purporting to entertain membership applications are perfunctory, to say the least. The reference in 2(2)(b) classifies certain schools as public buildings where they are not exempt from the operation of building regulations by virtue of section *4(1)(a)* of the *Building Act 1984*. Since the coming in to force of the *Education (Schools and Further and Higher Education) (Amendment) (England) Regulations 2000 (SI 2001/692)* on 1 April 2001, all schools in England are now covered by the regulations and are classified as public buildings. A similar situation has existed in Wales since 1 January 2002 following the passing of the *Education (Schools and Higher and Further Education)(Amendment)(Wales) Regulations 2001*.
(3) Any reference in these Regulations to a numbered regulation, Part or Schedule is a reference to the regulation, Part or Schedule so numbered in these Regulations.	

Building Regulations 2000	Commentary
PART II: CONTROL OF BUILDING WORK **3 Meaning of building work** (1) In these Regulations 'building work' means: (a) the erection or extension of a building;	Although the regulations do not attempt to define 'erection of a building' *section 123* of the *Building Act 1984* gives a relevant statutory definition for 'erection of a building' which includes *'the reconstruction of a building,* [and] *the roofing over of an open space between walls or buildings'*. More specifically, certain building operations are *'deemed to be the erection of a building'* as follows: *'(a) the re-erection of any building or part of a building when an outer wall has been pulled or burnt down to within ten feet of the surface of the ground adjoining the lowest storey of the building or of that part of the building,* *(b) the re-erection of a frame building or part of a frame building when that building or part of a building has been so far pulled down or burnt down, as to leave only the framework of the lowest storey of the building or of that part of the building,* (c) the roofing over of an open space between walls or buildings.'
(b) subject to paragraph (1A), the provision or extension of a controlled service or fitting in or in connection with a building;	*'controlled service or fitting'* is defined in regulation 2 above but see also the variation applied by virtue of regulation 3(1)(A) below in relation to work covered by paragraph L1 of Schedule 1.
(c) the material alteration of a building, or a controlled service or fitting, as mentioned in paragraph (2);	See the commentary note against *'material alteration'* in regulation 2(1) above and the note against regulation 3(2) below.
(d) work required by regulation 6 (requirements relating to a material change of use);	See note against regulation 6 below.
(e) the insertion of insulating material into the cavity wall of a building; (f) work involving the underpinning of a building.	The insertion of cavity wall insulation and the underpinning of a building are both controlled by the regulations. The insertion of cavity wall insulation must comply with the requirements of paragraph C4 (*Resistance to weather and ground moisture*) and paragraph D1 (*Cavity insulation*). With regard to underpinning, the regulations are applied so as to ensure that any movement of the building is stabilised. Regard will also need to be given to the effect on any sewers or drains near the work.

Building Regulations 2000	Commentary
(1A) The provision or extension of a controlled service or fitting: (a) in or in connection with an existing dwelling; and (b) being a service or fitting in relation to which paragraph L1, but not Part G, H, J or P of Schedule 1 imposes a requirement, shall only be building work where that work consists of the provision of a window, rooflight, roof window, door (being a door which together with its frame has more than 50% of its internal face area glazed), a space heating or hot water service boiler, or a hot water vessel.	The effect of this difficult wording is to place under control certain works in existing dwellings which are controlled only under paragraph L1. These are the provision of: • a window, rooflight, roof window or door (with (including its frame) more than 50% of its internal face area glazed), • a space heating or hot water service boiler, or • a hot water vessel. Any other works to which paragraph L1 applies (e.g. the installation of an energy-efficient lighting system in an existing dwelling) are not covered by the regulations.
(2) An alteration is material for the purposes of these Regulations if the work, or any part of it, would at any stage result: (a) in a building or controlled service or fitting not complying with a relevant requirement where previously it did; or in a building or controlled service or fitting which before the work commenced did not comply with a relevant requirement, being more unsatisfactory in relation to such a requirement. (3) In paragraph (2) 'relevant requirement' means any of the following applicable requirements of Schedule 1, namely: Part A (structure) paragraph B1 (means of warning and escape) paragraph B3 (internal fire spread structure) paragraph B4 (external fire spread) paragraph B5 (access and facilities for the fire service) Part M (access to and use of buildings)	Put more simply, an alteration is deemed to be a *material alteration* and so subject to control under the regulations if the alteration would cause the existing building to: (1) not meet certain specified requirements (the *relevant requirements*) of the regulations which it did meet before the alterations were commenced, or (2) be made worse in relation to compliance with the relevant requirements if the building was already not in compliance before the alterations were commenced. The *relevant requirements* are: • Part A (structure), • In Part B (Fire safety), paragraph B1 (means of warning and escape), paragraph B3 (internal fire spread structure), paragraph B4 (external fire spread) and paragraph B5 (access and facilities for the fire service), and • Part M (access to and use of buildings)

Building Regulations 2000	Commentary
	In general, it is not necessary to bring the existing building up to current regulation standards, however, it should not be made worse as a result of the works of alteration when measured against the relevant requirements. Special note should be made of the phrase '*at any stage*' in the first line of regulation 2(2) above, since this indicates that the regulations apply not only to completed work but also to partially completed work and, indeed, to work in progress. This has obvious implications for structural alterations (unsatisfactory temporary support) and alterations that affect a means of escape (where it might be blocked by the work of alteration).

4 Requirements relating to building work
(1) Building work shall be carried out so that:
  (a) it complies with the applicable requirements contained in Schedule 1; and
  (b) in complying with any such requirement there is no failure to comply with any other such requirement.
(2) Building work shall be carried out so that, after it has been completed:
  (a) any building which is extended or to which a material alteration is made; or
  (b) any building in, or in connection with, which a controlled service or fitting is provided, extended or materially altered; or
  (c) any controlled service or fitting;
 	
complies with the applicable requirements of Schedule 1 or, where it did not comply with any such requirement, is no more unsatisfactory in relation to that requirement than before the work was carried out.
	Regulation 4 makes it clear that building work (whether to a new or existing building) must comply with the applicable requirements of Schedule 1. These requirements are phrased in functional terms and cover such aspects of building design and construction as structural stability, safety in fire, resistance to dampness, ventilation, drainage, waste disposal, etc. They are given in full at the end of this commentary. It is possible that satisfying one part of Schedule 1 can cause a contravention of another part (e.g. from a purely structural viewpoint, external walls could be constructed of 190 millimetres thick masonry, however this form of construction would be unlikely to satisfy the weather resistance requirements of Part C and the heat loss requirements of Part L). Therefore, regulation 4(1)(b) prevents this by requiring that all parts of Schedule 1 be complied with (one requirement must not cause a contravention of another).
	  Work to existing buildings must be carried out so that, when the work is complete, the building is not made worse in relation to compliance with the regulations than it was before the work commenced.

Building Regulations 2000	Commentary
5 Meaning of material change of use For the purposes of paragraph 8(1)(e) of Schedule 1 to the Act and for the purposes of these Regulations, there is a material change of use where there is a change in the purposes for which or the circumstances in which a building is used, so that after that change: (a) the building is used as a dwelling, where previously it was not; (b) the building contains a flat, where previously it did not; (c) the building is used as an hotel or boarding house, where previously it was not; (d) the building is used as an institution, where previously it was not; (e) the building is used as a public building, where previously it was not; (f) the building is not a building described in Classes I to VI in Schedule 2, where previously it was; (g) the building, which contains at least one dwelling, contains a greater or lesser number of dwellings than it did previously; (h) the building contains a room for residential purposes, where previously it did not; (i) the building, which contains at least one room for residential purposes, contains a greater or lesser number of such rooms than it did previously, or (j) the building is used as a shop, where previously it was not.	The *Building Act, Schedule 1, paragraph 8(1)(e)* states that: *'Building regulations may be made with respect to:* *(e) buildings or parts of buildings, together with any services, fittings or equipment provided in or in connection with them, in cases where the purposes for which or the manner or circumstances in which a building or part of a building is used change or changes in a way that constitutes a material change of use of the building or part within the meaning of the expression "material change of use" as defined for the purposes of this paragraph by building regulations.'* Regulation 5 gives substance to this part of the Building Act and sets out details of the use changes which constitute building work under regulation 3(1)(d). Most of the building types referred to in regulation 5 are defined in regulation 2 above including dwelling, flat, institution, shop, public building and room for residential purposes. Reference should be made to the commentary to regulation 2 for more information on these. With regard to hotels and boarding houses (regulation 5(c)), the *Fire Precautions Act 1971* requires many of these premises to carry a fire certificate. Decided case law would indicate that the use of premises for bed-sitting accommodation is materially different from hotel use (by virtue of the fact that hotel populations are normally transient whereas bed-sitting rooms are usually people's homes).

Building Regulations 2000	Commentary
	Regulation 5(f) makes it a material change of use if a building was originally constructed so as to be exempt under Schedule 2 of the Building Regulations (Classes I to VI) and it is subsequently used for another purpose. Interestingly, Class VII (Extensions) is not included in this regulation. Therefore, it would appear that, for example, the conversion of a small-detached domestic garage to an office would constitute a material change of use, whereas the change of use of an attached conservatory to an office at the same property would not (although it might require planning approval). Regulations 5(h) and 5(i) were added in response to SI 2002/2871 which amended Part E (Resistance to the passage of sound) of the 2000 Regulations.
6 Requirements relating to material change of use (1) Where there is a material change of use of the whole of a building, such work, if any, shall be carried out as is necessary to ensure that the building complies with the applicable requirements of the following paragraphs of Schedule 1: (a) in all cases, B1 (means of warning and escape) B2 (internal fire spread – linings) B3 (internal fire spread – structure) B4(2) (external fire spread – roofs) B5 (access and facilities for the fire service) C2(c) (interstitial and surface condensation) F1 (ventilation) G1 (sanitary conveniences and washing facilities) G2 (bathrooms) H1 (foul water drainage) H6 (solid waste storage) J1 to J3 (combustion appliances)	Regulation 6 lists those parts of Schedule 1 which are applicable to the various types of material change of use described in regulation 5. The items listed in 6(1)(a) apply to all changes of use where the change affects the whole of a building. The items listed in 6(1)(b) to 6(1)(g) apply additional parts of Schedule 1 to specified use changes, for example Part E (Resistance to the passage of sound) only applies to changes which result in the creation of dwellings, flats, hotels and boarding houses and rooms for residential purposes. Where a change of use affects only part of a building, in most cases it is only that part which must comply with the listed parts of Schedule 1 (i.e. not the whole building). The exceptions are in relation to: • buildings exceeding 15 metres in height (see 6(1)(c) where the whole building must comply with B4(1) (External fire spread – walls);

Building Regulations 2000	Commentary
L1 (conservation of fuel and power – dwellings) L2 (conservation of fuel and power – buildings other than dwellings) P1 and P2 (electrical safety); (b) in the case of a material change of use described in regulations 5(c), (d), (e) or (f), A1 to A3 (structure); (c) in the case of a building exceeding 15 metres in height, B4(1) (external fire spread – walls); (cc) in the case of a material change of use described in regulations 5(a), (b), (c), (d), (g), (h), (i) or, where the material change provides new residential accommodation, (f), C1(2) resistance to contaminants; (d) in the case of a material change of use described in regulation 5(a), C2 (resistance to moisture); (e) in the case of a material change of use described in regulation 5(a), (b), (c), (g), (h) or (i), E1 to E3 (resistance to the passage of sound). (f) in the case of a material change of use described in regulation 5(e), where a public building consists of or contains a school, E4 (acoustic conditions in schools). (g) in the case of a material change of use described in regulation 5(c), (d), (e), or (j), M1 (access and use). (2) Where there is a material change of use of part only of a building, such work, if any, shall be carried out as is necessary to ensure that: (a) that part complies in all cases with any applicable requirement referred to in paragraph (1)(a); (b) in a case to which sub-paragraphs (b), (d), (e) or (f) of paragraph (1) apply, that part complies with the requirements referred to in the relevant sub-paragraph;	• case 6(1)(g) where M1 (access and use) is applied to the part being changed and to any sanitary conveniences provided in or in connection with the part being changed. In this case it should be noted that the whole building must comply with requirement M1(a) of Schedule 1 to the extent that reasonable provision is made to provide either suitable independent access to the part being changed or suitable access through the building to that part.

Building Regulations 2000	Commentary

(c) in a case to which sub-paragraph (c) of paragraph (1) applies, the whole building complies with the requirement referred to in that sub-paragraph; and
(d) in a case to which sub-paragraph (g) of paragraph (1) applies:
  (i) that part and any sanitary conveniences provided in or in connection with that part comply with the requirements referred to in that sub-paragraph; and
  (ii) the building complies with requirement M1(a) of Schedule 1 to the extent that reasonable provision is made to provide either suitable independent access to that part or suitable access through the building to that part.	

7 Materials and workmanship
Building work shall be carried out:
(a) with adequate and proper materials which:
  (i) are appropriate for the circumstances in which they are used,
  (ii) are adequately mixed and prepared, and
  (iii) are applied, used or fixed so as adequately to perform the functions for which they are designed; and
(b) in a workmanlike manner.
	This is a general statutory obligation imposed on the builder. Guidance on how the obligation may be met is contained in Approved Document to support regulation 7 (Materials and workmanship), although the guidance is of a very general nature. It should be noted that building regulations are concerned only with matters of health, safety welfare and convenience, conservation of fuel and power and access for disabled people. Therefore, for most parts of the regulations it is only against these criteria that the materials and workmanship may be judged (exceptions include Part E, paragraph H2 and paragraph J6). Building regulations do not guarantee that a building will be fit for its purpose or that building work will be good value for money and done to a high standard, etc.

Building Regulations 2000	Commentary
8 Limitation on requirements Parts A to D, F to K, N and P (except for paragraphs H2 and J6) of Schedule 1 shall not require anything to be done except for the purpose of securing reasonable standards of health and safety for persons in or about buildings (and any others who may be affected by buildings, or matters connected with buildings).	See also the note to regulation 7 above. From a reading of regulation 8 it is evident that the health and safety limitations do not apply to Part E (Resistance to the passage of sound), Part L (Conservation of fuel and power), Part M (Access and facilities for disabled people), requirement H2 (Wastewater treatment systems and cesspools) and requirement J6 (Protection against pollution). Parts E and M are excluded because they address the welfare and convenience of building users. Part L is excluded because it addresses conservation of fuel and power. Parts H2 and J6 are excluded because they deal directly with the prevention of contamination of water. All these matters are amongst the purposes, other than health and safety that may be addressed by building regulations under the *Building Act 1984, section 1(1).*
9 Exempt buildings and work Subject to paragraph (2) these Regulations do not apply to: (1) (a) the erection of any building or extension of a kind described in Schedule 2; or (b) the carrying out of any work to or in connection with such a building or extension, if after the carrying out of that work it is still a building or extension of a kind described in that Schedule. (2) The requirements of Part P of Schedule 1 apply to: (a) any greenhouse; and (b) any small detached building falling within Class VI in Schedule 2, which in either case receives its electricity from a source shared with or located inside the dwelling.	The regulations do not apply to the erection of any building set out in Classes I to VI of Schedule 2, or to extension work in Class VII. Furthermore, they have no application at all to any work done to or in connection with buildings in Classes I to VII provided that, in the case of Classes I to VI, the work does not involve a material change of use (see also note to regulation 5 above).

Building Regulations 2000	Commentary
PART III: EXEMPTION OF PUBLIC BODIES FROM PROCEDURAL REQUIREMENTS **10 The Metropolitan Police Authority** (1) The Metropolitan Police Authority is hereby prescribed for the purposes of section 5 of the Act (exemption of public bodies from the procedural requirements and enforcement of building regulations). (2) The Metropolitan Police Authority is exempt from compliance with these Regulations in so far as the requirements in these Regulations are not substantive requirements.	As a prescribed 'public body' under the provisions of the *Building Act 1984, section 5*, the Metropolitan Police Authority is exempt from compliance with the procedural requirements and enforcement provisions of the building regulations. Instead, the Metropolitan Police Authority is bound by the provisions of the *Building Act 1984, section 54* and by *Part VII* of the *Building (Approved Inspectors etc) Regulations 2000 (SI 2000/2532)* (as amended).
PART IV: RELAXATION OF REQUIREMENTS **11 Power to dispense with or relax requirements** (1) The powers under section 8(1) of the Act to dispense with or relax any requirement contained in these Regulations shall be exercisable by the local authority. (2) Any notification by the local authority to an applicant that they have refused his application to dispense with or relax any requirement of these Regulations shall inform the applicant of the effect of section 39(1) and (3) of the Act (appeal against refusal, etc. to relax building regulations).	The relaxation and dispensation of building regulations is covered in the *Building Act 1984, sections 8 to 11*, whereby the Secretary of State may give a direction dispensing with or relaxing a requirement if he considers that the operation of the requirement would be unreasonable in the circumstances of that particular case. This power has been delegated to local authorities by virtue of regulation 11(1) opposite. In practice, since the Building Regulations are phrased in functional terms with the intention of giving designers and builders flexibility in the way they comply, it may be very difficult for a local authority to selectively relax some aspects of a functional requirement, which is written in terms of reasonable standards of health and safety or other reasonable provisions. Additionally, if a decision is made to appeal against a refusal of a local authority to grant a relaxation or dispensation (regulation 11(2)), it is likely to be very difficult to argue the case unless very special circumstances exist.

Building Regulations 2000	Commentary
PART V: NOTICES AND PLANS **12 Giving of a building notice or deposit of plans** (1) In this regulation 'relevant use' means a use as a workplace of a kind to which Part II of the *Fire Precautions (Workplace) Regulations 1997* applies or a use designated under section 1 of the *Fire Precautions Act 1971*.	The *Fire Precautions (Workplace) Regulations 1997* apply to all workplaces including those covered by other fire-specific fire safety legislation (such as those for which a Fire Certificate is in force, or has been applied for under the *Fire Precautions Act 1971*). The *Fire Precautions Act 1971* requires a fire certificate to be in force for the following designated premises: • Hotels or boarding houses where sleeping accommodation is provided for more than six staff or guests (or some sleeping accommodation is provided above the first floor or below the ground floor); and • Factories, offices, shops and railway premises where more than 20 people are employed or more than 10 people work other than on the ground floor, or in factories only, where explosive or highly flammable materials are stored or used.
(2) Subject to the following provisions of this regulation, a person who intends to carry out building work or to make a material change of use shall: (a) give to the local authority a building notice in accordance with regulation 13; or (b) deposit full plans with the local authority in accordance with regulation 14.	Paragraph 12(2) only applies if the decision is taken to use the local authority. This requirement should be read in conjunction with regulation 20 below, since it is possible to use the services of an Approved Inspector as an alternative to a local authority for building regulation compliance. Where the local authority route is used two alternative methods of seeking approval are given in paragraph 12(2) – the service of a 'building notice' or the deposit of 'full plans'. These terms are defined in regulation 2(1) above. Although there are no 'prescribed forms' for a building notice or full plans application, all local authorities supply pre-printed forms for use by applicants.

Building Regulations 2000	Commentary
(3) A person shall deposit full plans where he intends to carry out building work in relation to a building put or intended to be put to a use which is a relevant use. (4) A person shall deposit full plans where he intends to carry out work which includes the erection of a building fronting on to a private street. (4A) A person shall deposit full plans where he intends to carry out building work in relation to which paragraph H4 of Schedule 1 imposes a requirement.	The use of a 'building notice' is severely restricted by virtue of paragraphs 12(3), 12(4) and 12(4A). In practice, a building notice may only be given where the work involves the construction, extension or material alteration of a dwelling, or the change of use to provide one or more dwellings. This option may be further reduced if: • The building (including a dwelling) fronts on to a private street (12(4)), or • The work involves construction over or adjacent to a drain, sewer or disposal main that is shown on the sewerage records of the sewerage undertaker (12(4A)).
(5) A person who intends to carry out building work is not required to give a building notice or deposit full plans where the work consists only of work: (a) described in column 1 of the Table in Schedule 2A if the work is to be carried out by a person described in the corresponding entry in column 2 of that Table, and paragraphs 1 and 2 of that Schedule have effect for the purposes of the descriptions in the Table; (b) Described in Schedule 2B.	It is not always necessary to notify the local authority or to use an Approved Inspector for certain types of building work, provided that the person carrying out the work is suitably qualified and experienced (see Schedule 2A). This is mainly concerned with work to install, alter or extend most *controlled services or fittings* (see definition and comment in regulation 2(1) above).
(6) Where regulation 20 of the Building (Approved Inspectors etc) Regulations 2000 (local authority powers in relation to partly completed work) applies, the owner shall comply with the requirements of that regulation instead of with this regulation.	Regulation 20 of the Building (Approved Inspectors etc) Regulations allows that where work is being supervised by an Approved Inspector and the initial notice ceases to be in force, unless another Approved Inspector can be appointed, the work will revert to the local authority. The local authority can then request certain plans of the work and may, under certain circumstances, require the opening up of already completed work to ascertain compliance with the regulations.

Building Regulations 2000	Commentary
13 Particulars and plans where a building notice is given (1) A building notice shall state the name and address of the person intending to carry out the work and shall be signed by him or on his behalf, and shall contain or be accompanied by: 　(a) a statement that is given for the purposes of regulation 12(2)(a); 　(b) a description of the proposed building work or material change of use; and 　(c) particulars of the location of the building to which the proposal relates and the use or intended use of that building. (2) In the case of the erection or extension of a building, a building notice shall be accompanied by: 　(a) a plan to a scale of not less than 1:1250 showing: 　　(i) the size and position of the building, or the building as extended, and its relationship to adjoining boundaries; 　　(ii) the boundaries of the curtilage of the building, or the building as extended, and the size, position and use of every other building or proposed building within that curtilage; 　　(iii) the width and position of any street on or within the boundaries of the curtilage of the building or the building as extended; 　(b) a statement specifying the number of storeys (each basement level being counted as one storey), in the building to which the proposal relates; and 　(c) particulars of: 　　(i) the provision to be made for the drainage of the building or extension; [(ii)] 　　(iii) the steps to be taken to comply with any local enactment which applies.	Regulation 13 gives details of the information and plans which are required to be submitted when a building notice is given to the local authority. The enables the local authority to identify the site, the building and its relationship with the site boundaries, the type of work being contemplated and the size and use of the building. Additionally the local authority will be able to ascertain what steps it needs to take to ensure that the building is being adequately drained and to what extent the building work is in compliance with any local Acts of Parliament. Where the work relates to the erection or extension of a building the building notice must be accompanied by a site plan to a scale of not less than 1:1250 which shows the building in relation to its boundaries, neighbouring buildings and streets, etc. Paragraph 13(2)(c)(ii) related to the precautions to be taken in building over a drain, sewer or disposal main. It was revoked by the *Building (Amendment) Regulations 2001* with effect from 1 April 2002 when section 18 of the Building Act was repealed and its provisions replaced by Requirement H4 of the building regulations.

Building Regulations 2000	Commentary
(3) In the case of building work which involves the insertion of insulating material into the cavity walls of a building, a building notice shall be accompanied by a statement which specifies: (a) the name and type of insulating material to be used; (b) the name of any European Technical Approval issuing body which has approved the insulating material; (c) the requirements of Schedule 1 in relation to which any body referred to in (b) has approved the insulating material; (d) any national standard of a member state of the European Economic Area to which the insulating material conforms; and (e) the name of any body which has issued any current approval to the installer of the insulating material. (4) Where building work involves the provision of a hot water storage system in relation to which paragraph G3 in Schedule 1 (hot water storage) imposes a requirement, a building notice shall be accompanied by a statement which specifies: (a) the name, make, model and type of hot water storage system to be installed; (b) the name of the body, if any, which has approved or certified that the system is capable of performing in a way which satisfies the requirements of paragraph G3 of Schedule 1; (c) the name of the body, if any, which has issued any current registered operative identity card to the installer or the proposed installer of the system.	Paragraphs 13(3) and 13(4) contain details of additional information which must accompany a building notice where there is an intention to insert insulating material into the cavity walls of a building (13(3)) or provide an unventilated hot water storage system in compliance with Paragraph G3 of Schedule 1 (13(4)). This is to ensure that the relevant work is of a suitable specification and is installed by a suitably qualified and experienced person.

Building Regulations 2000	Commentary
(5) Where a building notice has been given, a person carrying out building work or making a material change of use shall give the local authority, within such time as they specify, such plans as are, in the particular case, necessary for the discharge of their functions in relation to building regulations and are specified by them in writing.	The local authority is entitled to ask for sufficient plans (see comment to definition of 'full plans' in regulation 2(1) above) that are necessary for it to carry out its functions under building regulations. The authority must specify in writing the plans that are to be provided and it can impose time limits on the supply of such plans. Paragraph 13(5) should not be seen as a licence for the local authority to insist on 'full plans' being supplied by the developer since, as is seen in paragraph 13(6), the authority has no powers to pass or reject the plans supplied to accompany a building notice (the 'plans' are not deemed to have been deposited with the local authority and therefore cannot be passed or rejected under section 16 of the Building Act 1984). Clearly, the extent of the additional information that is requested will depend on the type and complexity of the work being carried out. For example, the construction of a new dwelling-house using the building notice procedure would undoubtedly result in a considerable amount of additional information being sought by the local authority. Since the developer, having produced this information, would not have the benefit of having his plans passed by the local authority it is probably more sensible to use the full plans procedure in such cases. The building notice procedure is most useful for minor alterations and extensions to dwellings, where the production of scale plans is not usually needed.
(6) Neither a building notice nor plans which accompany it or are given under paragraph (5) are to be treated for the purposes of section 16 of the Act as having been deposited in accordance with building regulations. (7) A building notice shall cease to have effect on the expiry of 3 years from the date on which that notice was given to the local authority, unless before the expiry of that period: (a) the building work to which the notice relates was commenced; or (b) the material change of use described in the notice was made.	It should be noted that a building notice automatically ceases to have effect 3 years from the date on which it was given to the local authority if the work has not started or the change of use has not taken place. This contrasts with the situation regarding the deposit of full plans. These also cease to have effect after 3 years (see the *Building Act 1984, section 32*) but the local authority must write to the owner and declare them to be of no effect for the provision to apply.

Building Regulations 2000	Commentary
14 Full plans (1) Full plans shall be accompanied by a statement that they are deposited for the purpose of regulation 12(2)(b). (2) (a) Full plans shall be deposited in duplicate, of which the local authority may retain one copy; and (b) where Part B of Schedule 1 (Fire safety) imposes a requirement in relation to proposed building work, an additional two copies of any such plans as demonstrate compliance with that requirement shall be deposited, both of which may be retained by the local authority.	In general, when using the local authority full plans route for compliance, it is necessary to deposit two copies of the plans. The local authority is entitled to retain one copy, and although most local authorities return the other copy endorsed with a pass or reject stamp as appropriate, there is no obligation to do so. The obligation under the *Building Act 1984, section 16*, is to give notice of pass or rejection to the person depositing the plans, not return the plans. Two additional copies of the plans must be deposited if Part B of Schedule 1 (Fire safety) applies to the intended building work. These additional plans are required so that the local authority can carry out its consultations with the fire authority and both copies may be retained. This requirement for extra copies does not apply where the proposed building work relates to the erection, extension or material alteration of a dwelling-house or flat.

(3) Full plans shall consist of:
  (a) a description of the proposed building work or material change of use, and the plans, particulars and statements required by paragraphs (1) to (4) of regulation 13; and
  (aa) where paragraph H4 of Schedule 1 imposes a requirement, particulars of the precautions to be taken in building over a drain, sewer or disposal main to comply with the requirements of that paragraph;
	Paragraph H4 of Schedule 1 relates to the erection or extension of a building, or to works of underpinning, adjacent to or over a drain, sewer or disposal main that is shown on the sewerage records of the sewerage undertaker. The work must be carried out in a way that is not detrimental to the building or extension, or to the continued maintenance of the drain, sewer or disposal main. See also regulation 14A below.

(b) any other plans which are necessary to show that the work would comply with these Regulations.	See the commentary to the definition of 'full plans' above for examples of 'other plans' referred to opposite.

Building Regulations 2000	Commentary
(4) Full plans shall be accompanied by a statement as to whether the building is put or is intended to be put to a use which is a relevant use as defined by regulation 12(1)	For details of what is regarded as a 'relevant use' see the commentary to regulation 12(1) above.
(5) Full plans may be accompanied by a request from the person carrying out building work that on completion of the work he wishes the local authority to issue a completion certificate in accordance with regulation 17. (6) Paragraph (2)(b) shall not require the deposit of additional copies of plans where the proposed building work relates to the erection, extension or material alteration of a dwelling-house or flat.	The full plans application form supplied by the local authority will usually contain a 'tick box' indicating that the applicant wishes the local authority to issue a completion certificate when the works are satisfactorily completed. This certificate will relate to all the applicable requirements of Schedule 1. Where a 'tick box' is not provided the applicant will have to formally request such a certificate. If this is not done the local authority will only be under an obligation to issue a completion certificate for work which is subject to the requirements of Part B of Schedule 1 (Fire safety). A completion certificate is a valuable document and will almost certainly be requested by the purchaser's solicitor if the building is subsequently sold.
14A Consultation with sewerage undertaker (1) This regulation applies where full plans have been deposited with the local authority and paragraph H4 of Schedule 1 imposes requirements in relation to the building work which is the subject of those plans.	Paragraph H4 of Schedule 1 relates to the erection or extension of a building, or to works of underpinning, adjacent to or over a drain, sewer or disposal main that is shown on the sewerage records of the sewerage undertaker.
(2) Where this regulation applies the local authority shall consult the sewerage undertaker: (a) as soon as practicable after the plans have been deposited; and (b) before issuing any completion certificate in relation to the building work in accordance with regulation 17 pursuant to a request under regulation 14(5).	Where Paragraph H4 of Schedule 1 applies, Regulation 14A requires that the local authority consults with the sewerage undertaker in the following cases: • before passing plans, and • before giving a completion certificate

Building Regulations 2000	Commentary
(3) Where a local authority is required by paragraph (2) to consult the sewerage undertaker they shall: (a) give to the sewerage undertaker, in a case where they are consulting them following the deposit of full plans, sufficient plans to show whether the work would, if carried out in accordance with those plans, comply with the applicable requirements of paragraph H4 of Schedule 1; (b) have regard to any views expressed by the sewerage undertaker; and (c) not pass plans or issue a completion certificate until 15 days have elapsed from the date on which they consulted the sewerage undertaker, unless the sewerage undertaker has expressed its views to them before the expiry of that period.	The consultation procedure (which involves the supply of plans and other information to the sewerage undertaker by the local authority) is necessary, to ensure that the sewerage authority's legal interests (involving access to and structural stability of sewers) are preserved. Further consultation may also be necessary if the actual construction work reveals considerable variation from that shown on the plans. The building notice procedure cannot be used where Paragraph H4 applies. A similar consultation process exists if an Approved Inspector is used under *regulation 13A* of the *Building (Approved Inspectors etc) Regulations 2000*.
15 Notice of commencement and completion of certain stages of work (1) subject to paragraph (8), a person who proposes to carry out building work shall not commence that work unless: (a) he has given the local authority notice that he intends to commence work; and (b) at least 2 days have elapsed since the end of the day on which he gave the notice. (2) subject to paragraph (8), a person carrying out building work shall not: (a) cover up any excavation for a foundation, any foundation, any damp-proof course or any concrete or other material laid over a site; or (b) cover up in any way any drain or sewer to which these Regulations apply, unless he has given the local authority notice that he intends to commence that work, and at least 1 day has elapsed since the end of the day on which he gave the notice.	Regulation 15 requires that a person who intends to carry out building work under local authority supervision must notify the local authority before work starts and before and after certain building operations are carried out. Notice must also be given on completion of the work and when the building (or a part of the building) is occupied if this occurs before completion. The time periods for the notices are given in regulation 15 and are quoted in days (see definition of 'day' in regulation 2(1) above). These requirements for the giving of notices apply only where a person is obliged under regulation 12 to give a building notice or deposit full plans, therefore they do not apply where an Approved Inspector is used or where the work is carried out by a person referred to in Schedule 2A.

Building Regulations 2000	Commentary
(3) subject to paragraph (8), a person who has laid, haunched or covered any drain or sewer in respect of which Part H of Schedule 1 (drainage and waste disposal) imposes a requirement shall give notice to that effect to the local authority not more than 5 days after the completion of the work. (4) subject to paragraph (8), a person carrying out building work shall, not more than 5 days after that work has been completed, give the local authority notice to that effect. (5) Where a building is being erected, and that building (or any part of it) is to be occupied before completion, the person carrying out the work shall give the local authority at least 5 days notice before the building or any part of it is occupied. (6) Where a person fails to comply with paragraphs (1) to (3), he shall comply within a reasonable time with any notice given by the local authority requiring him to cut into, lay open or pull down so much of the work as prevents them from ascertaining whether these Regulations have been complied with. (7) If the local authority have given notice specifying the manner in which any work contravenes the requirements in these Regulations, a person who has carried out any further work to secure compliance with these Regulations shall within a reasonable time after the completion of such further work give notice to the local authority of its completion. (8) Paragraphs (1) to (4) apply only to a person who is required by regulation 12 to give a notice or deposit full plans.	If a person fails to comply with the notice requirements of regulation 15, the local authority may require him by notice to cut into, lay open or pull down any work so that it can find out whether or not the regulations have been complied with. Where the local authority has served notice specifying that the work contravenes the regulations, on completion of the remedial work, notice must be given to the authority within a reasonable time. The term 'person carrying out building work' is not defined in the regulations, however, decided case law would indicate that this can mean the owner of a building who authorises a contractor to carry out building works on his behalf. Additionally, the *Building Act, section 36* enables the local authority to take action against the owner of the building in cases where the building regulations have been contravened.

Building Regulations 2000	Commentary
16 Energy rating (1) This regulation applies where a new dwelling is created by building work or by a material change of use in connection with which building work is carried out. (2) Where this regulation applies, the person carrying out the building work shall calculate the energy rating of the dwelling by means of a procedure approved by the Secretary of State, and shall give notice of that rating to the local authority. (3) The notice referred to in paragraph (2) shall be given not later than the date on which the notice required by paragraph (4) of regulation 15 is given, and, where a new dwelling is created by the erection of a building, it shall be given at least 5 days before occupation of the dwelling. (4) Where this regulation applies, subject to paragraphs (6) and (7), the person carrying out the building work shall affix, as soon as practicable, in a conspicuous place in the dwelling, a notice stating the energy rating of the dwelling. (5) The notice referred to in paragraph (4) shall be affixed not later than the date on which the notice required by paragraph (4) of regulation 15 is given, and, where a new dwelling is created by the erection of a building, it shall be affixed not later than 5 days before occupation of the dwelling.	Where a new dwelling is created, either by new building work or by a material change of use, the person carrying out the building work must calculate the energy rating of the dwelling by the government approved Standard Assessment Procedure (SAP) and must give the local authority notice of this within 5 days of completion of the dwelling or at least 5 days before occupation if this occurs before completion. A notice stating the energy rating for the dwelling must be affixed in a conspicuous place in the dwelling by the person carrying out the building work, unless he intends to occupy the dwelling himself. A similar notice procedure exists if an Approved Inspector is used under *regulation 12 of the Building (Approved Inspectors etc) Regulations 2000*.

Building Regulations 2000	Commentary
(6) Subject to paragraph (7), if, on the date the dwelling is first occupied as a residence, no notice has been affixed in the dwelling in accordance with paragraph (4), the person carrying out the building work shall, not later than the date on which the notice required by paragraph (4) of regulation 15 is given, give to the occupier of the dwelling a notice stating the energy rating of the dwelling calculated in accordance with paragraph (2). (7) Paragraphs (4) and (6) shall not apply in a case where the person carrying out the work intends to occupy, or occupies, the dwelling as a residence.	

16A Provisions applicable to self-certification schemes
(1) This regulation applies where building work consists only of work of a type described in column 1 of Schedule 2A and the work is carried out by a person who is described in column 2 of that Schedule in respect of that type of work.
(2) Where this regulation applies, the local authority is authorised to accept, as evidence that the requirements of regulations 4 and 7 have been satisfied, a certificate to that effect by the person carrying out the building work.
(3) Where this regulation applies, the person carrying out the work shall, not more than 30 days after completion of the work:
(a) give to the occupier a copy of the certificate referred to in paragraph (2); and
(b) give to the local authority:
  (i) notice to that effect; or
  (ii) the certificate referred to in paragraph (2)
(4) Paragraph 3 of this regulation does not apply where a person carries out the building work described in Schedule 2B which consists only of work on a low voltage or an extra-low voltage electrical installation.
	Regulation 16A makes special provision for building work listed in column 1 of the Table to Schedule 2A (see below). It authorises a local authority to accept, as evidence that the work complies with regulations 4 and 7, a certificate to that effect from a person listed in column 2 of Schedule 2A.
	
	It also provides for notification of the completion of the work where no certificate is to be given. Therefore, where a listed person carries out the work there is no need to submit plans or give a building notice to the local authority.
	
	Additionally, for certain types of work to low voltage or extra-low voltage installations described in Schedule 2B (see below) there is no need even to issue a certificate to the local authority or notify them of completion of the work. Such work is termed 'non-notifiable' however, whether carried out professionally or on a DIY basis, it must still comply with the requirements of Part P.

Building Regulations 2000	Commentary

17 Completion certificates
(1) A local authority shall give a completion certificate in accordance with this regulation and as provided for in paragraph (2) where:
(a) they receive a notice under regulation 15(4) or (5) that building work has been completed, or, that a building has been partly occupied before completion; and
(b) they have either
  (i) been notified in accordance with regulation 14(4) that the building is put or is intended to be put to a use which is a relevant use as defined by regulation 12(1); or
  (ii) been requested in accordance with regulation 14(5), to give a completion certificate.
(2) Where in relation to any building work or, as the case may be, to any part of a building which has been occupied before completion, a local authority have been able to ascertain, after taking all reasonable steps in that behalf, that the relevant requirements of Schedule 1 specified in the certificate have been satisfied, they shall give a certificate to that effect.
(3) In this regulation the relevant requirements mean:
(a) in a case mentioned in paragraph (1)(b)(i), the applicable requirements of Part B of Schedule 1 (fire safety);
(b) in a case mentioned in paragraph (1)(b)(ii), any applicable requirements of Schedule 1.
(4) A certificate given in accordance with this regulation shall be evidence (but not conclusive evidence) that the requirements specified in the certificate have been complied with.
	The full plans application form supplied by the local authority will usually contain a 'tick box' indicating that the applicant wishes the local authority to issue a completion certificate when the works are satisfactorily completed. This certificate will relate to all the applicable requirements of Schedule 1. Where a 'tick box' is not provided the applicant will have to formally request such a certificate. If this is not done the local authority will only be under an obligation to issue a completion certificate for work which is subject to the requirements of Part B of Schedule 1 (Fire safety). A completion certificate is a valuable document and will almost certainly be requested by the purchaser's solicitor if the building is subsequently sold. The local authority only becomes liable to issue a completion certificate when a notice is served on it under regulation 15 stating that the building work has been completed or that a building has been occupied (either in total or in part) before completion, and when the local authority is satisfied, after having taken all reasonable steps, that the relevant requirements of Schedule 1 specified in the certificate have been satisfied. It should be noted that a local authority cannot be fined for contravening this regulation (see regulation 22 below).

Building Regulations 2000	Commentary
PART VI: MISCELLANEOUS	
18 Testing of building work The local authority may make such tests of any building work as may be necessary to establish whether it complies with regulation 7 or any of the applicable requirements contained in Schedule 1. **19 Sampling of material** The local authority may take such samples of the material to be used in the carrying out of building work as may be necessary to enable them to ascertain whether such materials comply with the provisions of these Regulations.	Regulations 18 and 19 empower the local authority to test building work to ensure compliance with the requirements of regulation 7 (*materials and workmanship*) and any applicable parts of Schedule 1, and to take samples of materials to be used in the carrying out of the building work. The power to test all building work covered by the regulations was an extension of the former powers which applied only to the testing of drains and sewers. This necessitated amending regulation 18 (*see the Building (Amendment) Regulations 2001 SI 2001/3335*) and was needed so that local authorities could enforce the airtightness and insulation continuity requirements of Part L (*Conservation of fuel and power*) of Schedule 1. It should be noted (see regulation 20(2) below) that regulations 18 and 19 do not apply where the work is supervised by an Approved Inspector; however, similar powers exist in the *Building (Approved Inspectors etc) Regulations 2000* (as amended).
20 Supervision of building work otherwise than by local authorities (1) Regulations 12, 15, 16, [16A], 17, 18, 19 and 20A shall not apply in respect of any work specified in an initial notice, an amendment notice or a public body's notice, which is in force. (2) Regulations 18 and 19 shall not apply in respect of any work in relation to which a final certificate or a public body's final certificate has been accepted by the local authority.	As has been mentioned at intervals in the comments above, any kind of building work may be supervised by an Approved Inspector instead of by the local authority at the discretion of the person carrying out the building work. Approved Inspectors must comply with the *Building (Approved Inspectors etc) Regulations 2000* (as amended) and these are shown in Appendix 2 together with a commentary. Where an Approved Inspector supervises the work, certain regulations (which apply only to building work supervised by the local authority) will, of course, not apply. These are listed in regulation 20.

Building Regulations 2000	Commentary
20A (1) Subject to paragraph (4) below, this regulation applies to: (a) building work in relation to which paragraph E1 of Schedule 1 imposes a requirement; and (b) work which is required to be carried out to a building to ensure that it complies with paragraph E1 of Schedule 1 by virtue of regulation 6(1)(e) or 6(2)(b). (2) Where this regulation applies, the person carrying out the work shall, for the purpose of ensuring compliance with paragraph E1 of Schedule 1: (a) ensure that appropriate sound insulation testing is carried out in accordance with a procedure approved by the Secretary of State; and (b) give a copy of the results referred to in sub-paragraph (a) to the local authority. (3) The results of the testing referred to in paragraph (2)(b) shall be: (a) recorded in a manner approved by the Secretary of State; and (b) given to the local authority in accordance with paragraph (2)(b) not later than the date on which the notice required by regulation 15(4) is given. (4) Where building work consists of the erection of a dwelling-house or a building containing flats, this regulation does not apply to any part of the building in relation to which the person carrying out the building work notifies the local authority, not later than the date on which he gives notice of commencement of the work under regulation 15(1), that, for the purpose of achieving	Where it is proposed to erect dwelling-houses, flats, maisonettes or buildings containing rooms for residential purposes (see definition in regulation 2(1) above), or to change the use of a building to create these uses, regulation 20(A) requires that sound insulation testing (commonly known as pre-completion testing) be carried out on the structure (walls and floors) which forms the separation from other buildings and from other parts of the same building. The tests must be carried out according to a procedure approved by the Secretary of State and the results must be given to the local authority not more than 5 days after the building work has been completed. A similar procedure exists for the Approved Inspector control system. The requirement for sound insulation testing came into force on 1 July 2003 for buildings containing rooms for residential purposes. For houses, flats and maisonettes the requirement was deferred until 1 January 2004 to enable the house building industry to prepare a case for substituting the testing requirement with an alternative solution involving the adoption of robust standard details. Ultimately, this date was extended until 1 July 2004 when the new 'Robust Details' were ready to be launched. Therefore, there is an exemption from the need for pre-completion testing for people building new houses or buildings containing flats if they are using design details approved and published by Robust Details Ltd.

Building Regulations 2000	Commentary

compliance of the work with paragraph E1 of Schedule 1, he is using one or more design details approved by Robust Details Ltd, provided that:
(a) the notification specifies:
   (i) the part or parts of the building in respect of which he is using the design detail;
   (ii) the design detail concerned; and
   (iii) the unique number issued by Robust Details Ltd in respect of the specified use of that design detail;
(b) the building work carried out in respect of the part or parts of the building identified in the notification is in accordance with the design detail specified in the notification.	

21 Unauthorised building work (1) This regulation applies where it appears to a local authority that unauthorised building work has been carried out on or after 11 November 1985. (2) In this regulation, 'unauthorised building work' means building work other than work in relation to which an initial notice, an amendment notice or a public body's notice has effect, which is done without: (c) a building notice being given to the local authority; or (d) full plans of the work being deposited with the local authority; or (e) a notice of commencement of work being given, in accordance with regulation 15(1) of these Regulations, where a building notice has been given or full plans have been deposited.	Regulation 21 gives local authorities power to regularise unauthorised building work commenced on or after 11 November 1985 on application made by the building owner. Although there may be no legal imperative for the application (e.g. where the unauthorised work has been completed for more than 12 months) the procedure is often used where the building is to be sold and the local authority search reveals that approval was not sought for the development. This can hold up or prevent the sale. See also Section 6.2.1.

Building Regulations 2000	Commentary
(3) Where this regulation applies, the owner (in this regulation referred to as 'the applicant') may apply in writing to the local authority for a regularisation certificate in accordance with this regulation, and shall send with his application: (a) a statement that the application is made in accordance with this regulation, (b) a description of the unauthorised work, (c) so far as is reasonably practicable, a plan of the unauthorised work, and (d) so far as is reasonably practicable, a plan showing any additional work required to be carried out to secure that the unauthorised work complies with the requirements relating to building work in the building regulations which were applicable to that work when it was carried out (in this regulation referred to as 'the relevant requirements'). (4) Where a local authority receive an application in accordance with this regulation, they may require the applicant to take such reasonable steps, including laying open the unauthorised work for inspection by the authority, making tests and taking samples, as the authority think appropriate to ascertain what work, if any, is required to secure that the relevant requirements are met.	

Building Regulations 2000	Commentary
(5) When the applicant has taken any such steps required by the local authority as are described in paragraph (4), and having had regard to any direction given in accordance with sections 8 and 9 of, and Schedule 2 to, the Act dispensing with or relaxing a requirement in building regulations which applies to the unauthorised work the local authority shall notify the applicant: (a) of the work which in their opinion is required to comply with the relevant requirements or those requirements as dispensed with or relaxed, or (b) that they cannot determine what work is required to comply with the relevant requirements or those requirements as dispensed with or relaxed, or (c) that no work is required to secure compliance with the relevant requirements or those requirements as dispensed with or relaxed. (6) Where the local authority have been able to satisfy themselves, after taking all reasonable steps for that purpose that: (a) the relevant requirements have been satisfied (taking account of any work carried out and any dispensation or relaxation given in accordance with sections 8 and 9 of, and Schedule 2 to, the Act), or (b) no work is required to secure that the unauthorised work satisfies the relevant requirements (taking account of any such dispensation or relaxation), they may give a certificate to that effect (in this regulation referred to as 'a regularisation certificate').	

Building Regulations 2000	Commentary
(7) A regularisation certificate shall be evidence (but not conclusive evidence) that the relevant requirements specified in the certificate have been complied with. (8) Where this regulation applies, regulations 12 and 14 shall not apply, and neither the supply of plans nor the taking of any other action in accordance with this regulation is to be treated for the purposes of section 16 of the Act as the deposit of plans in accordance with building regulations.	
22 Contravention of certain regulations not to be an offence Regulation 17 is designated as a provision to which section 35 of the Act (penalty for contravening building regulations) does not apply.	See comment to regulation 17 above.

23 Transitional provisions
(1) Subject to paragraph (2), the Regulations specified in Schedule 3 shall continue to apply in relation to any building work as if these Regulations had not been made where:
  (a) before 1 January 2001 a building notice, an initial notice, an amendment notice or a public body's notice has been given to, or full plans have been deposited with, a local authority; and
  (b) building work is carried out or is to be carried out accordance with any such notice or plans, whether with or without any departure from such plans.
	See comment to regulation 1(1) above.

Building Regulations 2000	Commentary
(2) Where an initial notice given before 1 January 2001 is varied by an amendment notice given on or after that date, the Regulations specified in Schedule 3 shall continue to apply as if these Regulations had not been made, to so much of the building work as could have been carried out under that initial notice if the amendment notice had not been given. **24 Revocations** The Regulations specified in Schedule 3 are hereby revoked. *Nick Raynsford* Minister of State, Department of the Environment, Transport and the Regions 13 September 2000	

Building Regulations 2000	Commentary
SCHEDULE 1 REQUIREMENTS	The numbered regulations listed above deal with administrative and procedural issues that affect the way in which the building control system operates in England and Wales. These regulations put into effect the technical requirements (the 'substantive' provisions), which are contained in Schedule 1 in 14 Parts, arranged alphabetically from A to P (missing out the letters I and O). The requirements are expressed in broad functional terms, extensive use being made of words such as 'adequate', 'reasonable' and 'satisfactory'. This is deliberate, since the use of such language is designed to be compatible with the limitation on requirements described in regulation 8 above (i.e. for the purposes of securing reasonable standards of health and safety etc.). In theory, this allows designers the freedom to develop novel and innovative designs without being tied to prescriptive solutions, as was the case in the past. Guidance on what can be considered as adequate, reasonable or satisfactory may be found in a series of 'Approved Documents' approved by the Secretary of State for this purpose, each Part of Schedule 1 having a corresponding Approved Document (plus an extra one covering regulation 7 (Materials and workmanship). It should be noted that there is no obligation to use the Approved Document guidance; however, a designer whose solution varied from the guidance would have to demonstrate to the local authority or Approved Inspector that the Schedule 1 requirements had been met. The technical requirements of Schedule 1, together with a detailed study of the guidance options, are covered in detail in the other books in this series. Therefore, the comments to the Schedule 1 requirements included below relate only to procedural matters and do not deal with technical issues.

Building Regulations 2000	Commentary
SCHEDULE 1 REQUIREMENTS **Part A Structure** **Loading** **A1** (1) The building shall be constructed so that the combined dead, imposed and wind loads are sustained and transmitted by it to the ground: (a) safely; and (b) without causing such deflection or deformation of any part of the building, or such movement of the ground, as will impair the stability of any part of another building. (2) In assessing whether a building complies with sub-paragraph (1) regard shall be had to the imposed and wind loads to which it is likely to be subjected in the ordinary course of its use for the purpose for which it is intended. **Ground movement** **A2** The building shall be constructed so that ground movement caused by: (a) swelling, shrinkage or freezing of the subsoil; or (b) landslip or subsidence (other than subsidence arising from shrinkage), in so far as the risk can be reasonably foreseen, will not impair the stability of any part of the building. **Disproportionate collapse** **A3** The building shall be constructed so that in the event of an accident the building will not suffer collapse to an extent disproportionate to the cause.	Part A is quoted in regulation 3(3) as being one of the 'relevant requirements' which must be taken into account when judging whether or not an alteration is 'material' under the regulations and therefore subject to control. This ensures that alterations are carried out without putting the public at risk from structural failure both during the course of the work (by ensuring that adequate temporary support is provided) and on completion. It should be noted that whilst the underpinning of a building is clearly structural work, it is deemed to be 'building work' in its own right under regulation 3(1)(f). Part A is also applied to the material change of use of a building when the change involves the creation of: • a hotel or boarding house, • an institution, • a public building, and any building which is changed from an exempt use under Schedule 2 to a non-exempt use.

Building Regulations 2000	Commentary

Part B Fire Safety
 	
Means of warning and escape
B1 The building shall be designed and constructed so that there are appropriate provisions for the early warning of fire, and appropriate means of escape in case of fire from the building to place of safety outside the building capable of being safely and effectively used at all material times.
 	
Limits on application paragraph B1
Requirement B1 does not apply to any prison provided under section 33 of the Prisons Act 1952 (power to provide prisons, etc.).
 	
Internal fire spread (linings)
B2 (1) To inhibit the spread of fire within the building the internal linings shall:
    (a) adequately resist the spread of flame over their surfaces; and
    (b) have, if ignited, a rate of heat release which is reasonable in the circumstances.
  (2) In this paragraph 'internal linings' mean the materials lining any partition, wall, ceiling or other internal structure.
 	
Internal fire spread (structure)
B3 (1) The building shall be designed and constructed so that, in the event of a fire, its stability will be maintained for a reasonable period.
  (2) A wall common to two or more buildings shall be designed and constructed so that it adequately resists the spread of fire between those buildings. For the purposes of this sub-paragraph a house in a terrace and a semi-detached house are each to be treated as a separate building.
  (3) To inhibit the spread of fire within the building, it shall be sub-divided with fire-resisting construction to an extent appropriate to the size and intended use of the building.
	Requirements B1, B3, B4 and B5 of Part B are quoted in regulation 3(3) as being 'relevant requirements' which must be taken into account when judging whether or not an alteration is 'material' under the regulations and therefore subject to control. This ensures, for example, that when alterations are carried out the means of escape is not obstructed by the work and that access for the fire brigade is maintained should a fire occur during the course of the work. With the exception of requirement B4(1) External fire spread (walls), Part B applies to all classes of material change of use of a building. Requirement B4(1) applies only in the case of the change of use of a building which exceeds 15 metres in height and it applies to the whole of the building even where only part of the building has its use changed.

Building Regulations 2000	Commentary
Limits on application paragraph B3 Requirement B3(3) does not apply to material alterations to any prison provided under section 33 of the Prisons Act 1952. **B3** (4) The building shall be designed and constructed so that the unseen spread of fire and smoke within concealed spaces in its structure and fabric is inhibited. **External fire spread** **B4** (1) The external walls of the building shall adequately resist the spread of fire over the walls and from one building to another, having regard to the height, use and position of the building. (2) The roof of the building shall adequately resist the spread of fire over the roof and from one building to another, having regard to the use and position of the building. **Access and facilities for the fire service** **B5** (1) The building shall be designed and constructed so as to provide reasonable facilities to assist fire fighters in the protection of life. (2) Reasonable provision shall be made within the site of the building to enable fire appliances to gain access to the building.	
Part C Site preparation and resistance to contaminants and moisture **Preparation of site and resistance to contaminants** **C1** (1) The ground to be covered by the building shall be reasonably free from any material that might damage the building or affect its stability, including vegetable matter, topsoil and pre-existing foundations.	

Building Regulations 2000	Commentary
(2) Reasonable precautions shall be taken to avoid danger to health and safety caused by contaminants on or in the ground covered, or to be covered by the building and any land associated with the building. (3) Adequate sub-soil drainage shall be provided if it is needed to avoid: (a) the passage of ground moisture to the interior of the building; (b) damage to the building, including damage through the transport of water-borne contaminants to the foundations of the building. (4) For the purposes of this requirement, 'contaminant' means any substance, which is or may become harmful to any persons or buildings including substances, which are corrosive, explosive, flammable, radioactive or toxic.	
Resistance to moisture **C2** The walls, floors and roof of the building shall adequately protect the building and people who use the building from harmful effects caused by: (a) ground moisture; (b) precipitation including wind-driven spray; (c) interstitial and surface condensation; and (d) spillage of water from or associated with sanitary fittings or fixed appliances.	C2 is the only requirement in Part C which applies in the case of a material change of use of a building and then only where the change involves the creation of a dwelling (e.g. a barn conversion).

Building Regulations 2000	Commentary
Part D Toxic substances **Cavity insulation** **D1** If insulating material is inserted into a cavity in a cavity wall reasonable precautions shall be taken to prevent the subsequent permeation of any toxic fumes from that material into any part of the building occupied by people.	Part D has largely been rendered obsolete by the development of safe cavity fill insulating materials. It was originally brought in to control the use of urea formaldehyde foam insulants, which emitted toxic fumes when going through the process of curing. Such insulants are no longer used today, although the provisions of Part D would control the use of new insulants if they gave off noxious fumes.
Part E Resistance to the passage of sound **Protection against sound from other parts of the building and adjoining buildings** **E1** Dwelling-houses, flats and rooms for residential purposes shall be designed and constructed in such a way that they provide reasonable resistance to sound from other parts of the same building and from adjoining buildings. **Protection against sound within a dwelling-house, etc.** **E2** Dwelling-houses, flats and rooms for residential purposes shall be designed and constructed in such a way that: (a) internal walls between a bedroom or a room containing a water closet, and other rooms; and (b) internal floors, provide reasonable resistance to sound. ***Limits on application paragraph E2*** Requirement E2 does not apply to: (a) an internal wall which contains a door; (b) an internal wall which separates an en suite toilet from an associated bedroom; (c) existing walls and floors in a building which is subject to a material change of use.	Requirements E1, E2 and E3 now apply not only to dwellings but also to 'rooms for residential purposes'. This includes (but is not specifically limited to) rooms in hotels, hostels, boarding houses, halls of residence and residential homes. Specifically exempted from the definition are rooms in hospitals, or other similar establishments, used for patient accommodation. Obviously, the provision of sound insulation between rooms in such buildings might prevent staff from hearing if a patient was in distress and needed help. Requirements E1 to E3 also apply where a material change of use occurs involving the creation of a: • dwelling (or extra dwellings where one already exists) • flat • hotel or boarding house, or • room for residential purposes although the application of E2 is restricted to any new walls or floors provided by the works of conversion (i.e. it does not apply to existing walls and floors).

Building Regulations 2000	Commentary
Reverberation in common internal parts of buildings containing flats or rooms for residential purposes **E3** The common internal parts of buildings which contain flats or rooms for residential purposes shall be designed and constructed in such a way as to prevent more reverberation around the common parts than is reasonable. *Limits on application paragraph E3* Requirement E3 only applies to corridors, stairwells, hallways and entrance halls which give access to the flat or room for residential purposes.	
Acoustic conditions in schools **E4** (1) Each room or other space in a school building shall be designed and constructed in such a way that it has the acoustic conditions and the insulation against disturbance by noise appropriate to its intended use. (2) For the purposes of this Part 'school' has the same meaning as in section 4 of the Education Act 1996; and 'school building' means any building forming a school or part of a school.	Although schools are now subject to building regulation control this is likely to have little effect on the design and construction of such buildings since requirement E4 will continue to be met by complying with the values for sound insulation reverberation time and indoor ambient noise given in Section 1 of Building Bulletin 93 *The Acoustic Design of Schools* produced by the Department for Education and Science and published by the Stationery Office (ISBN 0 11 271105 7). E4 also applies where a public building containing a school is created by a material change of use.
Part F Ventilation	Compliance with requirement F1 has the added advantage of preventing action being taken against the building occupier under the *Workplace (Health, Safety and Welfare) Regulations 1992* (SI 1992/3004) when the building is in use. In mixed use developments the requirements of F1 for dwellings and for the non-domestic part of the use may differ. However, the requirements for the workplace use would apply to any shared parts of the building (such as common corridors and access stairways). This situation will also apply in flats, where it will still be necessary for people such as cleaners, wardens and caretakers to work in the common parts of the building.

Building Regulations 2000	Commentary
Means of ventilation **F1** There shall be adequate means of ventilation provided for people in the building. *Limits on application paragraph F1* Requirement F1 does not apply to a building or space within a building: (a) into which people do not normally go; or (b) which is used solely for storage; or (c) which is a garage used solely in connection with a single dwelling.	The purpose of the ventilation provisions in requirement F1 is to control moisture and pollutant levels in buildings so that they do not become a hazard to the health of the occupants. The requirement applies to all building types but excludes those where the provision of ventilation would not serve to protect the health of users.
Part G Hygiene **Sanitary conveniences and washing facilities** **G1** (1) Adequate sanitary conveniences shall be provided in rooms provided for that purpose, or in bathrooms. Any such room or bathroom shall be separated from places where food is prepared. (2) Adequate washbasins shall be provided in: (a) rooms containing water closets; or (b) rooms or spaces adjacent to rooms containing water closets. Any such room or space shall be separated from places where food is prepared. (3) There shall be a suitable installation for the provision of hot and cold water to washbasins provided in accordance with paragraph (2) (4) Sanitary conveniences and washbasins to which this paragraph applies shall be designed and installed so as to allow effective cleaning.	The requirement for rooms containing sanitary conveniences to be separated from places where food is prepared in G1 can be met by locating the sanitary conveniences in an adjoining room separated by a door. There is no need to provide a ventilated lobby between the two uses, as was previously the case.
Bathrooms **G2** A bathroom shall be provided containing either a fixed bath or shower bath, and there shall be a suitable installation for the provision of hot and cold water to the bath or shower bath.	

Building Regulations 2000	Commentary
Limits on application paragraph G2 Requirement G2 applies only to dwellings	The limit on application means that G2 applies only to houses, flats and maisonettes. Requirements G1 and G2 apply in all cases of material change of use.
Hot water storage **G3** A hot water storage system that has a hot water storage vessel which does not incorporate a vent pipe to the atmosphere shall be installed by a person competent to do so, and there shall be precautions: (a) to prevent the temperature of the stored water at any time exceeding 100°C; and (b) to ensure that the hot water discharged from safety devices is safely conveyed to where it is visible but will not cause danger to persons in or about the building. *Limits on application paragraph G3* Requirement G3 does not apply to: (a) a hot water storage system that has a storage vessel with a capacity of 15 litres or less; (b) a system providing space heating only; (c) a system which heats or stores water for the purposes only of an industrial process.	Requirement G3 is concerned with non-traditional hot water storage systems which are not vented to the atmosphere. Such systems are more efficient to run and do not require roofspace cold water storage tanks so are useful in flats. Because they are pressure sensitive there is the risk that unless adequate safety devices were installed a fault could lead to a steam explosion, with potentially devastating results, hence the precautions mentioned in sub-paragraphs (a) and (b) opposite. G3 applies to all building types within the limits on application and Part G is included within the definition of 'controlled service or fitting' in regulation 2(1). This means that any work involving the provision, extension or material alteration of an unvented hot water storage system is covered by the regulations.
Part H Drainage and waste disposal **Foul water drainage** **H1** (1) An adequate system of drainage shall be provided to carry foul water from appliances within the building to one of the following, listed in order of priority: (a) a public sewer; or, where that is not reasonably practicable, (b) a private sewer communicating with a public sewer; or, where that is not reasonably practicable,	Requirement H1 applies in all cases of material change of use and Part H is included within the definition of 'controlled service or fitting' in regulation 2(1). This means that any work involving the provision, extension or material alteration of a drainage and waste disposal system in a building is covered by the regulations. It should be noted that the preference is for foul water drainage to be carried to a public sewer (in contrast with the disposal of surface water, see comment to H3 below).

Building Regulations 2000	Commentary

(c) either a septic tank which has an appropriate form of secondary treatment or another wastewater treatment system; or, where that is not reasonably practicable,
    (d) a cesspool.
  (2) In this Part 'foul water' means waste water which comprises or includes:
    (a) waste from a sanitary convenience, bidet or appliance used for washing receptacles for foul waste; or
    (b) water which has been used for food preparation, cooking or washing.
 	
Limits on application paragraph H1
Requirement H1 does not apply to the diversion of water which has been used for personal washing or for the washing of clothes, linen or other articles to collection systems for reuse.	

Wastewater treatment systems and cesspools
H2 (1) Any septic tank and its form of secondary treatment, other wastewater treatment system or cesspool, shall be so sited and constructed that:
    (a) it is not prejudicial to the health of any person;
    (b) it will not contaminate any watercourse, underground water or water supply;
    (c) there are adequate means of access for emptying and maintenance; and
    (d) where relevant, it will function to a sufficient standard for the protection of health in the event of a power failure.	

Building Regulations 2000	Commentary
(2) Any septic tank, holding tank which is part of a wastewater treatment system or cesspool shall be: (a) of adequate capacity; (b) so constructed that it is impermeable to liquids; and (c) adequately ventilated. (3) Where a foul water drainage system from a building discharges to a septic tank, wastewater treatment system or cesspool, a durable notice shall be affixed in a suitable place in the building containing information on any continuing maintenance required to avoid risks to health.	The requirement for a maintenance notice in H2(3) is designed to ensure that occupants are aware that they are on a non-mains drainage system which requires regular emptying.
Rainwater drainage **H3** (1) Adequate provision shall be made for rainwater to be carried from the roof of the building. (2) Paved areas around the building shall be so constructed as to be adequately drained. *Limits on application paragraph H3(2)* Requirement H3(2) applies only to paved areas: (a) which provide access to the building pursuant to paragraph M1 (access and use), or requirement M2 (access to extensions to buildings other than dwellings); (b) which provide access to or from a place of storage pursuant to paragraph H6(2) of Schedule 1 (solid waste storage); or (c) in any passage giving access to the building, where this is intended to be used in common by the occupiers of one or more other buildings.	

Building Regulations 2000	Commentary
H3 (3) Rainwater from a system provided pursuant to sub-paragraphs (1) or (2) shall discharge to one of the following, listed in order of priority: (a) an adequate soakaway or some other adequate infiltration system; or, where that is not reasonably practicable, (b) a watercourse; or, where that is not reasonably practicable, (c) a sewer. *Limits on application paragraph H3(3)* Requirement H3(3) does not apply to the gathering of rainwater for reuse.	Requirement H3(3) makes it clear that the preferred method of disposal for rainwater drainage is to put it back into the ground in the vicinity of the building. This is to primarily to prevent overloading of sewerage systems and to lessen the incidence of flooding caused by taking rainwater to watercourses during spells of heavy rainfall.
Building over sewers **H4** (1) The erection or extension of a building or work involving the underpinning of a building shall be carried out in a way that is not detrimental to the building or building extension or to the continued maintenance of the drain, sewer or disposal main. (2) In this paragraph 'disposal main' means any pipe, tunnel or conduit used for the conveyance of effluent to or from a sewage disposal works, which is not a public sewer. (3) In this paragraph and paragraph H5 'map of sewers' means any records kept by a sewerage undertaker under section 199 of the Water Industry Act 1991. *Limits on application paragraph H4* Requirement H4 applies only to work carried out: (a) over a drain, sewer or disposal main which is shown on any map of sewers; or (b) on any site or in such a manner as may result in interference with the use of, or obstruction of the access of any person to, any drain, sewer or disposal main which is shown on any map of sewers.	Requirement H4 replaces section 18 of the *Building Act 1984* and may be administered by both local authorities and Approved Inspectors. The map of sewers referred to in H4(3) and in the limits on application (also in H5) means records kept by sewerage undertakers which show the location of sewers and contain descriptions of the effluent which may pass through them. In general, records are kept of public sewers (those which are adopted by the sewerage undertaker) and of disposal mains (pipes, tunnels or conduits used for the conveyance of effluent to or from a sewage disposal works which are not public sewers).

Building Regulations 2000	Commentary
Separate systems of drainage **H5** Any system for discharging water to a sewer which is provided pursuant to paragraph H3 shall be separate from that provided for the conveyance of foul water from the building. *Limits on application paragraph H5* Requirement H5 applies only to a system provided in connection with the erection or extension of a building where it is reasonably practicable for the system to discharge directly or indirectly to a sewer for the separate conveyance of surface water which is: (a) shown on a map of sewers; or (b) under construction either by the sewerage undertaker or by some other person (where the sewer is the subject of an agreement to make a declaration of vesting pursuant to section 104 of the Water Industry Act 1991).	The requirement of H5 for separate systems of drainage for foul and surface water is aimed at helping to minimise the volume of rainwater which enters the public foul sewer system since this can lead to overloading of the capacity of sewers and treatment works, and can cause flooding. Section 104 of the Water Industry Act 1991 enables a sewerage undertaker to agree in advance to adopt a sewer or sewage disposal works being constructed by a developer, provided that the work is completed to the authority's satisfaction. H5 duplicated some provisions of six local Acts of Parliament. Following consultation, these provisions were repealed on 1 March 2004 (see Section 4.3.2)
Solid waste storage **H6** (1) Adequate provision shall be made for storage of solid waste. (2) Adequate means of access shall be provided: (a) for people in the building to the place of storage; and (b) from the place of storage to a collection point (where one has been specified by the waste collection authority under section 46 (household waste) or section 47 (commercial waste) of the Environmental Protection Act 1990) or to a street (where no collection point has been specified).	Requirement H6 applies in all cases of material change of use. Sections 46 and 47 of the Environmental Protection Act 1990 allow the waste collection authority to specify the type and number of receptacles that should be provided and the position where the waste should be placed for collection.

Building Regulations 2000	Commentary
Part J Combustion appliances and fuel storage systems **Air supply** **J1** Combustion appliances shall be so installed that there is an adequate supply of air to them for combustion, to prevent overheating and for the efficient working of any flue. **Discharge of products of combustion** **J2** Combustion appliances shall have adequate provision for the discharge of products of combustion to the outside air. **Protection of building** **J3** Combustion appliances and flue pipes shall be so installed, and fireplaces and chimneys shall be so constructed and installed, as to reduce to a reasonable level the risk of people suffering burns or the building catching fire in consequence of their use. *Limits on application paragraphs J1, J2 and J3* Requirements J1, J2 and J3 apply only to fixed combustion appliances (including incinerators).	Requirements J1 to J3 apply in all cases of material change of use and Part J is included within the definition of 'controlled service or fitting' in regulation 2(1). This means that any work involving the provision, extension or material alteration of a combustion appliance or fuel storage system is covered by the regulations. This does not necessarily mean that the work must be supervised by a local authority or Approved Inspector. In cases where the scope of the work falls within the descriptions in column 1 of Schedule 2A (see below) and provided that the work is carried out by a person described in column 2 of that schedule, notification is not required.
Provision of information **J4** Where a hearth, fireplace, flue or chimney is provided or extended, a durable notice containing information on the performance capabilities of the hearth, fireplace, flue or chimney shall be affixed in a suitable place in the building for the purpose of enabling combustion appliances to be safely installed. **Protection of liquid fuel storage systems** **J5** Liquid fuel storage systems and the pipes connecting them to combustion appliances shall be so constructed and separated from buildings and the boundary of the premises as to reduce to a reasonable level the risk of the fuel igniting in the event of fire in adjacent buildings or premises.	Requirement J4 is intended to ensure that anyone carrying out work to a hearth, fireplace, flue or chimney is aware of the performance capabilities of the installation so that an appliance is not connected to an inappropriate flue, etc. It should be noted that the purpose of requirement J5 is to protect the liquid fuel storage system from a fire occurring within a building and not vice versa. Therefore, it applies to both oil and LPG storage installations. It will be noticed that there is no upper limit on the size of the installation described in the requirement and that it applies to all building types and all types of tanks (even those that are buried).

Building Regulations 2000	Commentary
Limits on application paragraph J5 Requirement J5 applies only to: (a) fixed oil storage tanks with capacities greater than 90 litres and connecting pipes; and (b) fixed liquefied petroleum gas storage installations with capacities greater than 150 litres and connecting pipes, which are located outside the building and which serve fixed combustion appliances (including incinerators) in the building.	
Protection against pollution **J6** Oil storage tanks and the pipes connecting them to combustion appliances shall: (a) be so constructed and protected as to reduce to a reasonable level the risk of the oil escaping and causing pollution; and (b) have affixed in a prominent position a durable notice containing information on how to respond to an oil escape so as to reduce to a reasonable level the risk of pollution. *Limits on application paragraph J6* Requirement J6 applies only to fixed oil storage tanks with capacities of 3,500 litres or less, and connecting pipes, which are: (a) located outside the building; and (b) serve fixed combustion appliances (including incinerators) in a building used wholly or mainly as a private dwelling, but does not apply to buried systems.	Requirement J6 is aimed at preventing the pollution of ground water should a spillage occur from an oil storage tank or from the pipes connecting it to a combustion appliance. The power to make such a regulation is contained in the *Building Act 1984* section 1(1)(c) and, along with requirement J5, it brings the regulations for England and Wales into line with those in Scotland regarding oil pollution and protection of liquid fuel storage systems.

Building Regulations 2000	Commentary
Part K Protection from falling, collision and impact	Compliance with the requirements in Part K has the added advantage of preventing action being taken against the building occupier under the *Workplace (Health, Safety and Welfare) Regulations 1992* (SI 1992/3004) when the building is in use. In mixed use developments the requirements of Part K for dwellings and for the non-domestic part of the use may differ (certain sections of Part K applying only to workplaces). However, the requirements for the workplace use would apply to any shared parts of the building (such as common corridors and access stairways). This situation will also apply in flats, where it will still be necessary for people such as cleaners, wardens and caretakers to work in the common parts of the building.
Stairs, ladders and ramps **K1** Stairs, ladders and ramps shall be so designed, constructed and installed as to be safe for people moving between different levels in or about the building. ***Limits on application paragraph K1*** Requirement K1 applies only to stairs, ladders and ramps which form part of the building.	One effect of the limit on application to both K1 and K2(a) is to include entrance steps situated on the outside of a building.
K2 (a) Any stairs, ramps, floors and balconies and any roof to which people have access, and (b) any light well, basement area or similar sunken area connected to a building, shall be provided with barriers where it is necessary to protect people in or about the building from falling. ***Limits on application paragraph K2(a)*** Requirement K2(a) applies only to stairs and ramps which form part of the building.	

Building Regulations 2000	Commentary

K3 (1) Vehicle ramps and any levels in a building to which vehicles have access, shall be provided with barriers where it is necessary to protect people in or about the building.
(2) Vehicle loading bays shall be constructed in such a way, or be provided with such features, as may be necessary to protect people in them from collision with vehicles.
 	
Protection from collision with open windows, etc.
K4 Provision shall be made to prevent people moving in or about the building from colliding with open windows, skylights or ventilators.
 	
Limits on application paragraph K4
Requirement K4 does not apply to dwellings.
 	
Protection against impact from and trapping by doors
K5 (1) Provision shall be made to prevent any door or gate:
    (a) which slides or opens upwards, from falling onto any person; and
    (b) which is powered, from trapping any person.
(2) Provision shall be made for powered doors and gates to be opened in the event of power failure.
(3) Provision shall be made to ensure a clear view of the space on either side of a swing door or gate.
 	
Limits on application paragraph K5
Requirement K5 does not apply to:
 	
(a) dwellings, or
(b) any door or gate which is part of a lift.	

Building Regulations 2000	Commentary

Part L Conservation of fuel and power
 	
Dwellings
L1 Reasonable provision shall be made for the conservation of fuel and power in dwellings by:
(a) limiting the heat loss:
  (i) through the fabric of the building;
  (ii) from hot water pipes and hot air ducts used for space heating;
  (iii) from hot water vessels;
(b) providing space heating and hot water systems which are energy efficient;
(c) providing lighting systems with appropriate lamps and sufficient controls so that energy can be used efficiently;
 	
Limits on application paragraph L1(c)
The requirement for sufficient controls in paragraph L1(c) applies only to external lighting systems fixed to the building.
 	
L1 (d) providing sufficient information with the heating and hot water services so that building occupiers can operate and maintain the services in such a manner as to use no more energy than is reasonable in the circumstances.
 	
Buildings other than dwellings
L2 Reasonable provision shall be made for the conservation of fuel and power in buildings other than dwellings by:
(a) limiting the heat losses and gains through the fabric of the building;
(b) limiting the heat loss:
  (i) from hot water pipes and hot air ducts used for space heating;
  (ii) from hot water vessels and hot water service pipes;
	Although the title of this Part (Conservation of fuel and power) has remained the same after the publication of the 2002 amendments, the emphasis in the guidance to Part L is clearly concerned with reducing CO_2 emissions. Therefore, the requirements are concerned not only with the design and construction of the external fabric of the building and its services but also with future use and maintenance of those services. Requirements L1 and L2 apply in all cases of material change of use and Part L is included within the definition of 'controlled service or fitting' in regulation 2(1). Effectively, the requirements apply to any work involving the provision, extension or material alteration of the services and fittings listed in requirements L1 and L2. In the case of work to existing dwellings affected by requirement L1, the need for compliance is restricted to the provision, extension or material alteration of:
	
	• a window, rooflight, roof window or door (with (including its frame) more than 50% of its internal face area glazed),
	• a space heating or hot water service boiler, or
	• a hot water vessel
	
	Any other works to which paragraph L1 applies (e.g. the installation of an energy-efficient lighting system in an existing dwelling) are not covered by the regulations.
	  It is not necessarily the case that the work must be supervised by a local authority or Approved Inspector. In cases where the scope of the work falls within the descriptions in column 1 of Schedule 2A (see below) and provided that the work is carried out by a person described in column 2 of that schedule, notification is not required.

Building Regulations 2000	Commentary
(c) providing space heating and hot water systems which are energy efficient; (d) limiting exposure to solar overheating; (e) making provision where air conditioning and mechanical ventilation systems are installed, so that no more energy needs to be used than is reasonable in the circumstances; (f) limiting the heat gains by chilled water and refrigerant vessels and pipes and air ducts that serve air conditioning systems; ***Limits on application paragraphs L2(e) and L2(f)*** Requirements L2(e) and (f) apply only within buildings and parts of buildings where more than 200 square metres of floor area is to be served by air conditioning or mechanical ventilation systems. **L2** (g) providing lighting systems which are energy efficient; ***Limits on application paragraphs L2(g)*** Requirement L2(g) applies only within buildings and parts of buildings where more than 100 square metres of floor area is to be served by artificial lighting. **L2** (h) providing sufficient information with the relevant services so that the building can be operated and maintained in such a manner as to use no more energy than is reasonable in the circumstances.	

Building Regulations 2000	Commentary
Part M Access to and use of buildings	The 2003 changes to Part M have resulted in a new title and a different emphasis, all references to disabled people having been dropped in the requirements. This brings the requirements closer to the sentiment expressed in the Building Standards (Scotland) Regulations. When considering access to and use of buildings attention must be drawn to other important legislation, most notably the *Disability Discrimination Act 1995* ('DDA'). This Act contains duties to make reasonable adjustments to physical features of premises in certain circumstances. Unfortunately, following the guidance in Approved Document M (which supports Part M) is not a requirement for satisfying these duties to make reasonable adjustments. However, the following points included in Approved Document M may prove helpful. **Duties in the Employment Field** From 1 October 2004, the exemption in the *Disability Discrimination (Employment) Regulations 1996* (SI1996/1456) whereby an employer was not required to alter any physical characteristic of a building if it complied with Part M of the Building Regulations ceases to apply. Some changes to the duty to make reasonable adjustments are introduced from 1 October 2004 and its coverage is extended to all employers (irrespective of size) and a range of other bodies and occupations. **Duties of providers of services to the public** From 1 October 2004: The duty to make reasonable adjustments set out in sections 21(2)(a), (b) and (c) of the DDA comes into force. It applies to all those who provide services to the public irrespective of their size. It requires service providers to take reasonable steps to remove, alter or provide a reasonable means of avoiding a physical feature of their premises, which makes it unreasonably difficult or impossible for disabled people to make use of their services.

Building Regulations 2000	Commentary
	Full details of the application of the DDA and of how it links with the Building Regulations may be found in Using the Building Regulations: Part M Access, ISBN 07506 64509 (*A Complete Guide to the Building Regulations for England and Wales, – Access to and use of buildings*).
	Part M is quoted in regulation 3(3) as being one of the 'relevant requirements' which must be taken into account when judging whether or not an alteration is 'material' under the regulations and therefore subject to control. This ensures that alterations are carried out without making buildings less accessible or usable than they were before the work was carried out (or during the course of the work). Requirement M1 also applies to the material change of use of a building where it is proposed to create a hotel or boarding house, institution, public building or shop. Where only part of such a building is having its use changed, the part changed and any sanitary conveniences provided in it, must comply with requirement M1. Additionally, suitable access to the part changed must be available either independently or through the building as a whole.
Access and use **M1** Reasonable provision shall be made for people to: (a) gain access to; and (b) use the building and its facilities. ***Limits on application Part M*** The requirements of this Part do not apply to: (a) an extension of or material alteration of a dwelling; or (b) any part of a building which is used solely to enable the building or any service or fitting in the building to be inspected, repaired or maintained.	Part M does not apply to an extension or material alteration of a dwelling, or to parts of any building which are concerned only with maintenance access.

Building Regulations 2000	Commentary
Access to extensions to buildings other than dwellings **M2** Suitable independent access shall be provided to the extension where reasonably practicable. *Limits on application paragraph M2* Requirement M2 does not apply where suitable access to the extension is provided through the building that is extended.	Taken together with the limits on application this means that independent access must be provided to a building extension where it is not possible to provide a suitable access through the existing building.
Sanitary conveniences in extensions to buildings other than dwellings **M3** If sanitary conveniences are provided in any building that is to be extended, reasonable provision shall be made within the extension for sanitary conveniences. *Limits on application paragraph M3* Requirement M3 does not apply where there is reasonable provision for sanitary conveniences elsewhere in the building, such that people occupied in, or otherwise having occasion to enter the extension, can gain access to and use those sanitary conveniences.	M3 would appear to insist on the provision of accessible sanitary conveniences in an extension to a building, if the original building already contained sanitary conveniences. This would be the case if it was not possible to gain access to the sanitary conveniences in the existing building from the extension. Where access is possible then the extension need not be fitted with accessible sanitary conveniences.
Sanitary conveniences in dwellings **M4** (1) Reasonable provision shall be made in the entrance storey for sanitary conveniences, or where the entrance storey contains no habitable rooms, reasonable provision for sanitary conveniences shall be made in either the entrance storey or principal storey. (2) In this paragraph 'entrance storey' means the storey which contains the principal entrance and 'principal storey' means the storey nearest to the entrance storey which contains a habitable room, or if there are two such storeys equally near, either such storey.	Although M4(1) seems fairly straightforward, the definitions provided in M4(2) make it very difficult to understand exactly what is required by this regulation. Put more simply, the intention of the regulation is to require accessible sanitary conveniences to be situated in the entrance storey of the dwelling, *if this contains any habitable rooms*. Otherwise, the sanitary conveniences may be placed in either the entrance storey or the nearest storey to the principal entrance, which contains habitable rooms.

Building Regulations 2000	Commentary
Part N Glazing – Safety in relation to impact, opening and cleaning **Protection against impact** N1 Glazing, with which people are likely to come into contact whilst moving in or about the building, shall: (a) if broken on impact, break in a way which is unlikely to cause injury; or (b) resist impact without breaking; or (c) be shielded or protected from impact.	Although N1 appears somewhat vague in extent (*glazing, with which people are likely to come into contact*) it is clarified in the guidance as meaning glazing in critical locations such as doors and door side panels, and the lower parts of glazed walls and partitions.
Manifestation of glazing N2 Transparent glazing, with which people are likely to come into contact while moving in or about the building, shall incorporate features which make it apparent. *Limits on application paragraph N2* Requirement N2 does not apply to dwellings.	N2 applies only to buildings other than dwellings and is aimed at reducing the risk of people colliding with transparent glazed areas which might otherwise not be apparent.
Safe opening and closing of windows, etc. N3 Windows, skylights and ventilators which can be opened by people in or about the building shall be so constructed or equipped that they may be opened, closed or adjusted safely. *Limits on application paragraph N3* Requirement N3 does not apply to dwellings. **Safe access for cleaning windows, etc.** N4 Provision shall be made for any windows, skylights, or any transparent or translucent walls, ceilings or roofs to be safely accessible for cleaning. *Limits on application paragraph N4* Requirement N4 does not apply to: (a) dwellings, or (b) any transparent or translucent elements whose surfaces are not intended to be cleaned.	Compliance with the requirements in N3 and N4 has the added advantage of preventing action being taken against the building occupier under the *Workplace (Health, Safety and Welfare) Regulations 1992* (SI 1992/3004) when the building is in use. (See also the comments under Part K above regarding common parts of flats and similar buildings).

Building Regulations 2000	Commentary
Part P Electrical safety **Design, installation, inspection and testing** **P1** Reasonable provision shall be made in the design, installation, inspection and testing of electrical installations in order to protect persons from fire or injury. **Provision of information** **P2** Sufficient information shall be provided so that persons wishing to operate, maintain or alter an electrical installation can do so with reasonable safety. *Limits on application paragraph Part P* The requirements in this Part apply only to electrical installations that are intended to operate at low or extra-low voltage and are: (a) in a dwelling; (b) in the common parts of a building serving one or more dwellings, but excluding power supplies to lifts; (c) in a building that receives its electricity from a source located within or shared with a dwelling; and (d) in a garden or in or on land associated with a building where the electricity is from a source located within or shared with a dwelling.	The requirements of Part P apply to electrical installations (or parts of electrical installations) that are intended to operate at low or extra-low voltage in buildings or parts of buildings comprising: • dwelling houses and flats • dwellings and business premises that have a common supply, for example shops and public houses with a flat above • common access areas in blocks of flats such as corridors and staircases (but not to the power supply to lifts) • shared amenities of blocks of flats such as laundries and gymnasia • in or on land associated with a building, for example Part P applies to fixed lighting and pond pumps in gardens • in outbuildings such as sheds, detached garages and greenhouses.

Building Regulations 2000	Commentary

SCHEDULE 2 – EXEMPT BUILDINGS AND WORK   Regulation 9
 	
CLASS I
 	
Buildings controlled under other legislation
(1) Any building the construction of which is subject to the Explosives Acts 1875 and 1923(a).
(2) Any building (other than a building containing a dwelling or a building used for office or canteen accommodation) erected on a site in respect of which a licence under the Nuclear Installations Act 1965(b) is for the time being in force.
(3) Any building included in the schedule of monuments maintained under section 1 of the Ancient Monuments and Archaeological Areas Act 1979(c).
 	
CLASS II
 	
Buildings not frequented by people
A detached building:
(a) into which people do not normally go; or
(b) into which people go only intermittently and then only for the purpose of inspecting or maintaining fixed plant or machinery,
unless any point of such a building is less than one and a half times its height from
  (i) any point of a building into which people can or do normally go; or
  (ii) the nearest point of the boundary of the curtilage of that building, whichever is the nearer.
	The regulations do not apply to the erection of any building set out in Classes I to VI of Schedule 2, or to extension work in Class VII. Furthermore, they have no application at all to any work done to or in connection with buildings in Classes I to VII provided that, in the case of Classes I to VI, the work does not involve a material change of use. Regulation 5(f) makes it a material change of use if a building was originally constructed so as to be exempt under Schedule 2 (Classes I to VI) and it is subsequently used for another purpose. Interestingly, Class VII (Extensions) is not included in this regulation. Therefore, it would appear that, for example, the conversion of a small-detached domestic garage to an office would constitute a material change of use, whereas the change of use of an attached conservatory to an office at the same property would not (although it might require planning approval). It should be noted that in the case of Class VII conservatories or porches, in order for the exemption to apply the glazing must satisfy the requirements of Part N (Glazing – safety in relation to impact, opening and cleaning). This raises and interesting conundrum since exempt work is not notifiable to the Local Authority or Approved Inspector, therefore no check would be carried out to confirm that the glazing did comply. However, it does make enforcement action possible where a breach of building regulations can be shown.
	
	This also applies in the case of the requirements of Part P (Electrical Safety), since the requirements of this Part apply to greenhouses and small-detached buildings in Class VI of Schedule 2, provided that the relevant building receives its electricity from a source shared with or located inside a dwelling and some minor work described in Schedule 2B is non-notifiable. Interestingly, proposals to bring conservatories back under control are currently out for consultation.

Building Regulations 2000	Commentary

CLASS III
 	
Greenhouses and agricultural buildings
(1) Subject to paragraph 3, a greenhouse.
(2) A building used, subject to paragraph 3, for agriculture, or a building principally for the keeping of animals, provided in each case that:
   (a) no part of the building is used as a dwelling;
   (b) no point of the building is less than one and a half times its height from any point of a building which contains sleeping accommodation; and
   (c) the building is provided with a fire exit which is not more than 30 metres from any point in the building.
(3) The descriptions of buildings in paragraphs 1 and 2 do not include a greenhouse or a building used for agriculture if the principal purpose for which they are used is retailing, packing or exhibiting.
(4) In paragraph 2, 'agriculture' includes horticulture, fruit growing, the growing of plants for seed and fish farming.
 	
CLASS IV
 	
Temporary buildings
A building which is not intended to remain where it is erected for more than 28 days.
 	
CLASS V
 	
Ancillary buildings
(1) A building on a site, being a building which is intended to be used only in connection with the disposal of buildings or building plots on that site.
(2) A building on the site of construction or civil engineering works, which is intended to be used only during the course of those works and contains no sleeping accommodation.	

Building Regulations 2000	Commentary

(3) A building, other than a building containing a dwelling or used as an office or showroom, erected for use on the site of and in connection with a mine or quarry.
 	
CLASS VI
 	
Small detached buildings
(1) A detached single storey building, having a floor area which does not exceed 30 square metres, which contains no sleeping accommodation and is a building:
  (a) no point of which is less than 1 metre from the boundary of its curtilage; or
  (b) which is constructed substantially of non-combustible material.
(2) A detached building designed and intended to shelter people from the effects of nuclear, chemical or conventional weapons, and not used for any other purpose, if:
  (a) its floor area does not exceed 30 square metres; and
  (b) the excavation for the building is no closer to any exposed part of another building or structure than a distance equal to the depth of the excavation plus 1 metre.
(3) A detached building, having a floor area which does not exceed 15 square metres, which contains no sleeping accommodation.
 	
CLASS VII
 	
Extensions
 	
The extension of a building by the addition at ground level of:
 	
(a) a conservatory, porch, covered yard or covered way; or
(b) a carport open on at least two sides;
 	
where the floor area of that extension does not exceed 30 square metres, provided that in the case of a conservatory or porch which is wholly or partly glazed, the glazing satisfies the requirements of Part N of Schedule 1.	

Schedule 2A

Exemptions from requirement to give building notice or deposit full plans

Regulation 12(5)

Column 1	Column 2
Type of work	**Person carrying out work**
Installation of a heat-producing gas appliance.	A person, or an employee of a person, who is a member of a class of persons approved in accordance with regulation 3 of the Gas Safety (Installation and Use) Regulations 1998
Installation of: (a) an oil-fired combustion appliance which has a rated heat output of 45 kilowatts or less and which is installed in a building with no more than three storeys (excluding any basement); or (b) oil storage tanks and the pipes connecting them to combustion appliances.	An individual registered under the Oil Firing Registration Scheme by the Oil Firing Technical Association for the Petroleum Industry Ltd in respect of that type of work.
Installation of a solid fuel burning combustion appliance which has a rated heat output of 50 kilowatts or less and which is installed in a building with no more than three storeys (excluding any basement).	An individual registered under the Registration Scheme for Companies and Engineers involved in the Installation and Maintenance of Domestic Solid Fuel Fired Equipment by HETAS Ltd in respect of that type of work.
Installation, as a replacement, of a window, rooflight, roof window or door in an existing building.	A person registered under the Fenestration Self-Assessment Scheme by Fensa Ltd in respect of that type of work.
Installation of fixed low or extra-low voltage electrical installations.	A person registered by BRE Certification Limited, British Standards Institution, ELECSA Limited, NOCEIC Certification Services Limited, or Zurich Certification Limited in respect of that type of work.
Any building work which is necessary to ensure that any appliance, service or fitting which is installed and which is described in the preceding entries in column 1 above, complies with the applicable requirements contained in Schedule 1.	The person who installs the appliance, service or fitting to which the building work relates and who is described in the corresponding entry in column 2 above.

Interpretation and application

(1) For the purposes of this Schedule:
'appliance' includes any fittings or services, other than a hot water storage vessel which does not incorporate a vent pipe to the atmosphere, which form part of the space heating or hot water system served by the combustion appliance; and 'building work' does not include the provision of a masonry chimney.

(2) The final entry in the table above does not apply to building work which is necessary to ensure that a heat-producing gas appliance complies with the applicable requirements contained in Schedule 1 unless the appliance:
(a) has a net rated heat input of 70 kilowatts or less; and
(b) is installed in a building with no more than three storeys (excluding any basement).

Commentary: Schedule 2A was inserted into the regulations by virtue of the *Building (Amendment) Regulations 2002 (SI 2002/440)* and it came into force on 1 April 2002. Its effect is to exempt certain kinds of work (described in column 1) from the need to give a building notice or deposit full plans (see regulation 12(5) above) provided that the work is carried out by a person described in column 2 of the Schedule. It relates mainly to the installation of services and fittings (including heating and hot water boilers and replacement doors and windows) when these are carried out by suitably qualified and experienced people.

Schedule 2B

Descriptions of work where no building notice or deposit of full plans is required

Regulation 12(5)
(1)   Work consisting of:
(a) replacing any socket outlet, control switch or ceiling rose;
(b) replacing a damaged cable for a single circuit only;
(c) re-fixing or replacing enclosures or existing installation components, where the circuit protective measures are unaffected;
(d) providing mechanical protection to an existing fixed installation, where the circuit protective measures and current carrying capacity of the conductors are unaffected by the increased thermal insulation.
(2)   Work which:
(a) is not in a kitchen, or a special location,
(b) does not involve work on a special installation, and
(c) consists of:
(i) adding light fittings and switches to an existing circuit;
(ii) adding socket outlets and fused spurs to an existing ring or radial circuit; or
(iii) installing or upgrading main or supplementary equipotential bonding.

(3) In paragraph 2:

'special installation' means an electric floor or ceiling heating system, a garden lighting or electric power installation, an electricity generator, or an extra-low voltage lighting system which is not a pre-assembled lighting set bearing the CE marking referred to in regulation 9 of the Electrical Equipment (Safety) Regulations 1994; and

'special location' means a location within the limits of the relevant zones specified for a bath, a shower, a swimming or paddling pool or a hot air sauna in the Wiring Regulations, sixteenth edition, published by the Institution of Electrical Engineers and the British Standards Institution as BS 7671: 2001 and incorporating amendments 1 and 2.

Commentary: Schedule 2B was inserted into the regulations by virtue of the *Building (Amendment) Regulations 2004 (SI 2004/1808)* and it came into force on 1 January 2005. Its effect is to exempt certain kinds of work to electrical installations covered by Part P of Schedule 1 from the need to give a building notice or deposit full plans (see regulation 12(5) above) or to issue a certificate of conformance or to notify the local authority on completion. In order for the work to be regarded as 'non-notifiable' under Schedule 2B it should not include the provision of a new circuit. Non-notifiable work includes most work on existing circuits, but not additions to existing circuits that are:

- located in kitchens or 'special locations' such as bathrooms and saunas, or
- associated with 'special installations' such as garden lighting systems and solar photovoltaic power supplies.

Schedule 3

Revocations of regulations

Regulation 24

Title	Reference
The Building Regulations 1991	SI 1991/2768
The Building Regulations (Amendment) Regulations 1992	SI 1992/1180
The Building Regulations (Amendment) Regulations 1994	SI 1994/1850
The Building Regulations (Amendment) Regulations 1995	SI 1995/1356
The Building Regulations (Amendment) Regulations 1997	SI 1997/1904
The Building Regulations (Amendment) Regulations 1998	SI 1998/2561
The Building Regulations (Amendment) Regulations 1999	SI 1999/77
The Building Regulations (Amendment) (No. 2) Regulations 1999	SI 1999/3410
The Building Regulations (Amendment) Regulations 2000	SI 2000/1554

Commentary on the Building (Approved Inspectors etc) Regulations 2000 (as amended)

Introduction

In the following commentary, the requirements of the *Building (Approved Inspectors etc) Regulations 2000 (SI 2000/2532)* have been combined with subsequent amendments (*SI 2001/3336; SI 2002/2872, SI 2003/3133 and SI 2004/1466*) to produce a consolidated version. This will provide the reader with a complete updated text and will avoid the need for constant cross-referencing amongst the documents listed above.

The notes in this commentary should be read in conjunction with the main text:

Building (Approved Inspectors etc) Regulations 2000	Commentary
Building (Approved Inspectors etc) Regulations 2000 (SI 2000 No. 2532) *Made*: 13 September 2000 *Laid before Parliament*: 22 September 2000 *Coming into force*: 1 January 2001 The Secretary of State, in exercise of the powers conferred upon him by sections 1(1), 16(9), 17(1) and (6), 35, 47(1) to (5), 49(1) and (5), 50(1), (4), (6) and (7), 51(1) and (2), 52(1) to (3) and (5), 53(2) and (4), 54(1) to (3) and (5), and 56(1) and (2) of, and Schedules 1 and 4 to, the Building Act 1984, and of all other powers enabling him in that behalf, after consulting the Building Regulations Advisory Committee and such other bodies as appear to him to be representative of the interests concerned in accordance with section 14(3) of that Act, hereby makes the following Regulations: **PART I: GENERAL** **1 Citation, commencement and revocations** (1) These Regulations may be cited as the Building (Approved Inspectors etc) Regulations 2000 and shall come into force on 1 January 2001 on which date the regulations specified in Schedule 1 shall be revoked.	The *Building (Approved Inspectors etc) Regulations 2000* replaced the *Building (Approved Inspectors etc) Regulations 1985 (SI 1985/1066)* consolidated all subsequent amendments to those regulations. However, the 1985 Regulations will continue to apply in relation to work which was notified to a local authority via an Approved Inspector's initial notice or amendment notice, or a public body's notice before the 1 January 2001. The various amending regulations have also been subject to certain transitional arrangements. The term 'Principal Regulations' contained in the text below means the *Building Regulations 2000*.
2 Interpretation	The interpretation section should be read in conjunction with the *Building Act 1984, section 126* (General interpretation) and the *Interpretation Act 1978* (which applies generally to all Acts of Parliament unless a contrary intention is expressed in another Act). The definitions contained in this section apply throughout the regulations unless specifically excluded elsewhere.

Building (Approved Inspectors etc) Regulations 2000	Commentary
(1) In these Regulations unless the context otherwise requires: 'the Act' means the Building Act 1984;	The *Building Act 1984* contains the power to make building regulations and many administrative provisions (including provisions concerned with the Approved Inspector system of control), which govern the running of the building control system in England and Wales. All references to 'the Act' in the text below are to the *Building Act 1984*.
'building' means any permanent or temporary building but not any other kind of structure or erection, and a reference to a building includes a reference to part of a building;	This definition of 'building' is also contained in the *Building Regulations 2000* and is narrower than that contained in the *Building Act 1984*, the purpose being to restrict the scope of the Regulations to what are commonly thought of and referred to as buildings. The Building Act definition is necessarily couched in very wide terms so as to provide the local authority with the necessary powers to deal with (e.g. dangerous structures and demolitions). In the past, doubt has been cast over the status of such structures as residential park homes and marquees. Provided that a residential park home conforms to the definition given in the *Caravan Sites and Control of Development Act 1960* (as augmented by the *Caravan Sites Act 1968*) it is exempt from the definition of 'building' contained in the regulations, and according to the *Manual to the Building Regulations* a marquee is not regarded as a building.
'building work' has the meaning given in regulation 3(1) of the Principal Regulations;	The Principal Regulations apply only to 'building work' as defined in regulation 3(1). Even so, certain buildings are exempt from the requirements of the regulations (such as those occupied by the Crown, and those belonging to Statutory Undertakers, the UK Atomic Energy Authority and the Civil Aviation Authority) so the definition does not apply to such buildings.

Building (Approved Inspectors etc) Regulations 2000	Commentary
'controlled service or fitting' means a service or fitting in relation to which Part G, H, J, or L of Schedule 1 to the Principal Regulations imposes a requirement;	This means the installation of bathroom fittings, hot water storage systems, sanitary conveniences, drainage and waste disposal systems, combustion appliances, and replacement doors, windows and rooflights, space heating and hot water boilers and vessels. This definition is different to that contained in regulation 2 of the Principal Regulations in that there is no reference to Part P Electrical Safety.
'day' means any period of 24 hours commencing at midnight and excludes any Saturday, Sunday, Bank holiday or public holiday;	The definition of 'day' is needed with respect to the periods of notice required by virtue of regulations 8, 9, 12, 12A, 13, 13A, 15, 17, 22, 25, 27 and 30.
'dwelling' includes a dwelling-house and a flat;	Although this does not actually define what a dwelling is, decided case law would indicate that the dwelling does not have to be occupied to be so classified. Additionally, it has been held that cooking facilities do not have to be available for the premises to come within the definition of 'dwelling'.
'dwelling-house' does not include a flat or a building containing a flat;	This definition clearly delineates houses from flats and allows separate provisions to be made for each.
'energy rating' of a dwelling means a numerical indication of the overall energy efficiency of that dwelling obtained by the application of a procedure approved by the Secretary of State under regulation 16(2) of the Principal Regulations;	Regulation 16(2) of the Principal Regulations requires that when a person carries out building work which involves the creation of a new dwelling (dwelling-house, flat or maisonette) he must calculate the energy rating of that dwelling in accordance with a standard procedure approved by the Secretary of State and described in Approved Document L1. Notice of the energy rating of the dwelling must be given to the Approved Inspector at the times indicated in regulation 12.
'fire authority' means the authority discharging in the area in which the premises are or are to be situated the functions of fire authority under the Fire Services Act 1947;	Under the Fire Services Act 1947, fire authorities are county councils of non-metropolitan counties; metropolitan county and civil defence authorities in the six metropolitan counties and the London Fire and Civil Defence Authority in Greater London.

Building (Approved Inspectors etc) Regulations 2000	Commentary
'flat' means separate and self-contained premises constructed or adapted for use for residential purposes and forming part of a building from some other part of which it is divided horizontally;	This definition covers maisonettes as well as flats. The main problem with this definition lies in the interpretation of the phrase *separate and self-contained premises*. In the past questions have arisen over the status of staff flats (e.g. over licensed premises) where these are used exclusively by the pub manager or his staff, particularly in relation to the application of the resistance to the passage of sound requirements of Part E to the Principal Regulations. Where such 'flats' are used for residential purposes and they are separately approached and have separate lockable front doors it is reasonable to assume that Part E does apply and that the premises described are truly flats within the definition. Decided case law is relatively quiet on the subject of *separate and self-contained premises*; however in the case of *Malekshed v Howarde Walden Estates Ltd, Times Law Reports, 9 July 2001*, it was held that a substantial terraced house in Central London physically linked by a basement to a mews property at the rear could reasonably be called a house, although in this case both properties had been demised together under the terms of a Head Lease.
'material alteration' has the meaning given in regulation 3(2) of the Principal Regulations;	Building work which constitutes of a 'material alteration' to a building is subject to control under the Principal Regulations. The definition in regulation 3(2) is phrased in general terms and can cause problems of interpretation. The *Manual to the Building Regulations* makes it clear that works of repair (replacement, redecoration, routine maintenance and making good) do not come under the definition.

Building (Approved Inspectors etc) Regulations 2000	Commentary
'material change of use' has the meaning given in regulation 5 of the Principal Regulations;	Nine cases of material change of use are given in regulation 5 of the Principal Regulations and these differ substantially from the definition given under planning law in the *Town and Country Planning Act 1990, section 55*, so approval under the Principal Regulations may be required for a change of use where planning approval is not required and vice versa.
'the Principal Regulations' means the Building Regulations 2000.	
(2) Where any regulation requires the use of a numbered form in Schedule 2, a form substantially to the like effect may be used. (3) Any reference in these Regulations to the carrying out of work includes a reference to the making of a material change of use. (4) Any reference in these Regulations to an initial notice (whether or not combined with a plans certificate) shall in an appropriate case be construed as a reference to that initial notice as amended by an amendment notice which has been accepted by a local authority.	Schedule 2 contains the 'outline' of 12 prescribed forms and lists the information that must be included in each. The effect of regulation 2(2) is to allow Approved Inspectors to tailor the forms to their own 'house' style provided that the minimum information given in Schedule 2 is included in the relevant form. Several examples of such forms are given in Chapters 5 and 6.

Building (Approved Inspectors etc) Regulations 2000	Commentary
PART II: GRANT AND WITHDRAWAL OF APPROVAL **3 Approval of inspectors** (1) Where the Secretary of State has designated a body in accordance with regulation 4 (referred to in these Regulations as a 'designated body'), a person seeking to be an Approved Inspector shall apply to a designated body giving particulars of: (a) in the case of a person other than a body corporate, his qualifications and experience; and (b) in the case of a body corporate, the number, qualifications and experience of the people to be employed in the discharge of its functions under these Regulations, and the person shall answer any inquiries which that designated body makes about those matters. (2) Where there is no designated body, a person seeking to be an Approved Inspector shall apply to the Secretary of State giving particulars of: (a) in the case of a person other than a body corporate, his qualifications and experience; and (b) in the case of a body corporate, the number, qualifications and experience of the people to be employed in the discharge of its functions under these Regulations, and the person shall answer any inquiries which the Secretary of State makes about those matters.	A private individual or corporate body wishing to carry out building control functions as an Approved Inspector must be registered with the Construction Industry Council (CIC) under rules laid down by the Secretary of State. In order to be registered as an Approved Inspector a number of criteria must be met. These include the holding of suitable professional qualifications, demonstration of adequate practical experience and the carrying of suitable indemnity insurance. In accordance with the responsibilities entailed by CIC's appointment as the body designated to register Approved Inspectors, it established the Construction Industry Council Approved Inspectors Register (CICAIR) to maintain and operate the Approved Inspector Register. CICAIR provides applicants with a route to qualification as an Approved Inspector and, upon qualification, full Registration facilities.

Building (Approved Inspectors etc) Regulations 2000	Commentary
4 Designation of bodies to approve inspectors If it appears to the Secretary of State that a body might properly be designated as a body to approve inspectors he may, if the body consents, designate it for that purpose.	See note to regulation 3 above.
5 Manner of approval or designation The approval of an inspector or the designation of a body to approve inspectors shall be given to that person or body by a notice in writing specifying any limitation on the approval or designation.	

6 Termination of approval or designation
(1) The approval of an inspector given by a designated body or by the Secretary of State shall cease to have effect at the end of a period of 5 years from the date on which it was given.
(2) The approval of an inspector may be withdrawn by a notice in writing given to the inspector by the person who approved him.
(3) The Secretary of State may withdraw the designation of a designated body by giving the body notice in writing, but:
  (i) such withdrawal shall not affect the operation of any subsisting approval given by the body, and
  (ii) a subsisting approval may be withdrawn by the Secretary of State as if it had been given by him.
(4) Where an Approved Inspector is convicted of an offence under section 57 of the Act (false or misleading notices and certificates etc), the person by whom the approval was given may on receipt of a certificate of the conviction forthwith withdraw the approval and no further approval shall be given to an Approved Inspector whose approval has been withdrawn for a period of 5 years beginning with the date of his conviction.
	If an individual or corporate body is successful in its application to become an Approved Inspector, the approval will be for a period of 5 years. Further terms of approval may be sought.
	
	The approval of an inspector may be withdrawn by the approving body (the CIC) by notice in writing to him for example, if the inspector has contravened any relevant rules of conduct or has shown that he or she is unfitted for the work.
	
	More seriously, where an Approved Inspector is convicted of an offence under section 57 of the Act (which deals with false or misleading notices and certificates etc) the CIC may withdraw its approval. In this case the convicted person's name would be removed from the list for a period of 5 years. There is no provision for appeals or reinstatement. Additionally, the Secretary of State has powers to withdraw the designation of the approving body. If this were to happen it would not affect the status of any approvals given by the designated body, although the Secretary of State retains the right to withdraw an approval if he deems it necessary.

Building (Approved Inspectors etc) Regulations 2000	Commentary

7 Lists of approvals and designations
(1) The Secretary of State shall maintain:
   (a) a list of bodies which are for the time being designated by him for the purpose of approving inspectors, and
   (b) a list of inspectors for the time being approved by him.
(2) The Secretary of State shall:
   (a) supply to every local authority in whose area these Regulations apply a copy of the first lists of Approved Inspectors and designated bodies prepared by him under this regulation; and
   (b) notify every such local authority as soon as practicable of the withdrawal of any approval or designation and of any addition to the lists.
(3) A designated body shall:
   (a) maintain a list of inspectors for the time being approved by it; and
   (b) notify every local authority in whose area these Regulations apply as soon as practicable after withdrawing approval from any inspector.
(4) Lists maintained under this regulation shall set out any limitation placed on the approval or designation of the persons or bodies listed and shall indicate the date on which each approval will expire.
	Provision is made for the Secretary of State to keep lists of designated bodies and inspectors approved by him, and for their supply to local authorities. He must also keep lists up to date (if there are withdrawals or additions to the list) and must notify local authorities of these changes.
	   In the first years of the Approved Inspector system the Secretary of State approved all corporate bodies that wished to become Approved Inspectors. On 8 July 1996 the CIC was designated as the body responsible for approving non-corporate Approved Inspectors, although initially the Secretary of State reserved the right to continue to approve corporate bodies. From 1 March 1999 the CIC became responsible also for the approval of corporate Approved Inspectors.
	   The CIC must also maintain a list of inspectors that it has approved. There is no express provision for these lists to be open to public inspection, although the CIC is bound to inform the appropriate local authority if it withdraws its approval from any inspector.
	   In approving any inspector, either the Secretary of State or the designated body may limit the description of work in relation to which the person or body is approved. Any limitations will be noted in the official lists, as will any date of expiry of approval.

Building (Approved Inspectors etc) Regulations 2000	Commentary

PART III: SUPERVISION OF WORK BY APPROVED INSPECTORS
8 Initial notice
(1) The prescribed form of an initial notice:
   (a) which is not combined with a plans certificate, shall be Form 1 in Schedule 2; or
   (b) which is combined with a plans certificate, shall be Form 4 in Schedule 2.
(2) An initial notice shall be accompanied by the plans and documents described in the relevant form prescribed by paragraph (1).
(3) The grounds on which a local authority shall reject an initial notice are those prescribed in Schedule 3.
(4) The period within which a local authority may give notice of rejection of an initial notice is 5 days beginning with the day on which the notice is given.
	See sections 47 and 50 of the *Building Act 1984*.
	   When the client has appointed an Approved Inspector for a particular project, the first step in the Approved Inspector process will be the service of an initial notice on the local authority in whose area the work is to be carried out. This will make the local authority aware that building work in their area is being legally controlled under the Regulations, and will notify them of certain linked powers that they have under the *Building Act 1984* and any local Acts of Parliament. The local authority has 5 days in which to accept or reject the initial notice. A typical initial notice is illustrated in Figure 5.3.
	   The local authority's grounds for rejecting an initial notice are given in Section 5.6.5.

Building (Approved Inspectors etc) Regulations 2000	Commentary
9 Amendment notice (1) The prescribed form of an amendment notice shall be Form 2 in Schedule 2. (2) An amendment notice shall be accompanied by the plans and documents described in the form prescribed by paragraph (1). (3) The grounds on which a local authority shall reject an amendment notice are those prescribed in paragraphs 1 to 11 of Schedule 3. (4) The period within which a local authority may give notice of rejection of an amendment notice is 5 days beginning with the day on which the notice is given.	See section 51A of the *Building Act 1984*. Where an initial notice has been given and accepted for a scheme and it is proposed subsequently to make extensive design changes that materially alter the scope of the works it may be necessary for the person who is carrying out the work and the Approved Inspector jointly to give an amendment notice to the local authority. There is a prescribed form for an amendment notice and it must contain the information which is required for an initial notice (see Section 5.6.5) plus either: • a statement to the effect that all plans submitted with the original notice remain unchanged; or • copies of all the amended plans with a statement that any plans not included remain unchanged. The local authority has 5 working days in which to accept or reject the notice and it may only reject it on prescribed grounds. The procedure is identical to that for acceptance or rejection of an initial notice (see Section 5.6.5). Where copies of amended plans must be supplied, it is assumed that these are similar in content to those submitted with the original initial notice; that is where the work relates to the erection or extension of a building, a site plan to a scale of not less than 1:1250 which shows the boundaries and location of the site. There would be little point in submitting detailed plans of the revised design since the local authority would have no powers to inspect these and would not have been supplied with the original design details in any case. However, there is no definition of 'plans' in the Approved Inspector regulations.

Building (Approved Inspectors etc) Regulations 2000	Commentary
10 Independence of Approved Inspectors (1) For the purposes of this regulation 'minor work' means: (a) the material alteration or extension of a dwelling-house which before the work is carried out has two storeys or less and which afterwards has no more than three storeys; or (b) the provision, extension or material alteration of a controlled service or fitting in or in connection with any building; or (c) work consisting of the underpinning of a building; and for the purposes of this paragraph a basement is not to be regarded as a storey. (2) An Approved Inspector shall have no professional or financial interest in the work he supervises unless it is minor work. (3) A person shall be regarded as having a professional or financial interest in the work described in any notice or certificate given under these Regulations if: (a) he is or has been responsible for the design or construction of any of the work in any capacity, or (b) he or any nominee of his is a member, officer or employee of a company or other body which has a professional or financial interest in the work, or (c) he is a partner or is in the employment of a person who has a professional or financial interest in the work. (4) For the purposes of this regulation: (a) a person shall be treated as having a professional or financial interest in the work even if he has that interest only as trustee for the benefit of some other person, (b) in the case of married people living together, the interest of one spouse shall, if known to the other, be deemed to be also an interest of the other.	An Approved Inspector must have no professional or financial interest in the work supervised, unless it is minor work. For example, this prevents an Approved Inspector from supervising work that he or she has designed or constructed, or from working for a company which has an interest in the work. Minor work includes: • the extension or alteration of a single or two-storey dwelling-house (or its controlled services or fittings), provided that the dwelling-house does not exceed three storeys after the alterations (any basements may be disregarded when counting storeys). The definition of dwelling-house does not include flats; and • work involving the underpinning of a building. Independence is not required of an inspector supervising minor work but the limitation on the number of storeys should be noted. It should be noted that involvement in the work as an Approved Inspector on a fee basis is not a debarring interest!

Building (Approved Inspectors etc) Regulations 2000	Commentary
(5) For the purposes of this regulation: (a) involvement in the work as an Approved Inspector, (b) entitlement to any fee paid for his function as an Approved Inspector, and (c) potential liability to pay any sum if a claim is made under the insurance cover provided for the purposes of the Act, shall not be regarded as constituting a professional or financial interest.	
11 Functions of Approved Inspectors (1) Subject to paragraph (2), an Approved Inspector by whom an initial notice has been given shall, so long as the notice continues in force, take such steps (which may include the making of tests of building work and the taking of samples of material) as are reasonable to enable him to be satisfied within the limits of professional skill and care that: (a) regulations 4, 6 and 7 of the Principal Regulations are complied with, (b) subparagraph omitted (see commentary), (c) the requirements of regulation 12 and 12A of these Regulations are complied with. (2) In a case where any requirement of Part L of Schedule 1 to the Principal Regulations is to be complied with by the insertion of insulating material into the cavity in a wall after that wall has been constructed, the Approved Inspector need not supervise the insertion of the insulating material but shall state in the final certificate whether or not at the date of that certificate the material has been inserted. (3) Paragraph omitted (see commentary).	(1)(b) and (3) were omitted by virtue of the Building (Approved Inspectors etc) (Amendment) Regulations 2001 (SI 2001/3336). The functions of the Approved Inspector in regulation 11 to ensure compliance with the Principal Regulations are phrased in general terms (unlike the duties of the local authority to enforce the regulations in the Act). The general nature of regulation 11 makes it unclear as to the extent to which plans must be inspected or site visits carried out; however, it is likely that an Approved Inspector is liable for negligence and it is suggested, therefore, that he must inspect the work to ensure compliance, in contrast to local authorities who have a discretion as to whether to inspect or not. These issues are discussed in full in Section 5.8. In addition to the generalised functions discussed above the Approved Inspector must also ensure that: • the regulations concerning the calculation of energy ratings for dwellings, are complied with; • the regulation concerning sound-insulation testing, is complied with. • where building work involves the insertion of insulating material into the cavity in a wall after that wall has been built, the Approved Inspector is not required to supervise the insertion of the material, but must state in the final certificate (see Section 5.8.4) whether or not the material has been inserted.

Building (Approved Inspectors etc) Regulations 2000	Commentary

12 Energy rating
(1) This regulation applies where a new dwelling is created by building work or by a material change of use in connection with which building work is carried out, and the building work in question is the subject of an initial notice.
(2) Where this regulation applies, the person carrying out the building work shall calculate the energy rating of the dwelling and give notice of that rating to the Approved Inspector who gave the initial notice.
(3) The notice referred to in paragraph (2) shall be given not later than 5 days after completion of the dwelling, and
  (a) where the dwelling is created by building work, and
    (i) the dwelling is occupied, and
    (ii) no final certificate is given the notice shall be given not later than the end of the period of 8 weeks beginning with the date of occupation; or
  (b) where the dwelling is created by a material change of use in connection with which building work is carried out, and
    (i) the change of use takes place, and
    (ii) no final certificate is given the notice shall be given not later than the end of the period of 8 weeks beginning with the date on which the change of use takes place.
(4) Where this regulation applies, subject to paragraphs (6) and (8), the person carrying out the building work shall affix, as soon as practicable, in a conspicuous place in the dwelling, a notice stating the energy rating of the dwelling calculated in accordance with paragraph (2).
	Where a new dwelling is created by building work or by a material change of use, the client (i.e. the person carrying out the building work) is under an obligation to calculate the energy rating of the dwelling and give notice of it to the Approved Inspector. This must be done not later than 5 days after completion of the dwelling where it is created by building work or, if it is occupied before completion, not later than 8 weeks from the date of occupation.
	Where the dwelling is created by a material change of use, the notice must be given not later than 8 weeks from the date on which the change of use took place.
	Additionally, the person carrying out the building work is obliged either to display the notice of energy rating in a conspicuous place in the dwelling or to give a copy of it to the occupier, although this does not apply where the person carrying out the building work is also the occupier (i.e. for self-builders). Failure to provide the Approved Inspector with the energy rating notice, or to display it in the dwelling (or give it to the occupier, as appropriate), could result in the Approved Inspector being unable to give the final certificate. This could result in the consequences described in Section 5.8.4.

Building (Approved Inspectors etc) Regulations 2000	Commentary
(5) The notice referred to in paragraph (4) shall be affixed not later than 5 days after completion of the dwelling, and, in a case where subparagraph (b) of paragraph (3) applies, not later than the period of 8 weeks beginning with the date on which the change of use takes place. (6) Subject to paragraph (8), if, on the date the dwelling is first occupied as a residence, no notice has been affixed in the dwelling in accordance with paragraph (4), the person carrying out the building work shall give to the occupier of the dwelling a notice stating the energy rating of the dwelling calculated in accordance with paragraph (2). (7) The notice referred to in paragraph (6) shall be given not later than 5 days after completion of the dwelling, and, in a case where subparagraph (a) of paragraph (3) applies, not later than the end of the period of 8 weeks beginning with the date of occupation of the dwelling. (8) Paragraphs (4) and (6) shall not apply in a case where the person carrying out the work intends to occupy, or occupies, the dwelling as a residence.	
12A Sound insulation testing (1) Subject to paragraph (4) below this regulation applies to: (a) building work in relation to which paragraph E1 of Schedule 1 to the Principal Regulations imposes a requirement; and (b) work which is required to be carried out to a building to ensure that it complies with paragraph El of Schedule 1 to the Principal Regulations by virtue of regulation 6(1)(e) or 6(2)(b) of those Regulations, which is the subject of an initial notice.	An obligation is placed on the Approved Inspector (in the case of dwelling-houses, flats and rooms used for residential purposes) to check that appropriate sound insulation testing has been carried out by the developer to ensure compliance with Regulation E1 *(Protection against sound from other parts of the building and adjoining buildings)*. The results of the test must be recorded in a manner approved by the Secretary of State (see paragraph 1.41 of Approved Document E) and must be given to the Approved Inspector not later than 5 days after completion of the works.

Building (Approved Inspectors etc) Regulations 2000	Commentary
(2) Where this regulation applies, the person carrying out the work shall, for the purpose of ensuring compliance with paragraph E1 of Schedule 1: (a) ensure that appropriate sound-insulation testing is carried out in accordance with a procedure approved by the Secretary of State; and (b) give a copy of the results of the testing referred to in subparagraph (a) to the Approved Inspector who gave the initial notice. (3) The results of the testing referred to in paragraph (2)(a) shall be: (a) recorded in a manner approved by the Secretary of State; and (b) given to the Approved Inspector in accordance with paragraph (2)(b) not later than 5 days after completion of the work to which the initial notice relates. (4) Where the building work consists of the erection of a dwelling-house or a building containing flats, this regulation does not apply to any part of the building in relation to which the person carrying out the building work notifies the Approved Inspector prior to commencement of the building work on site that, for the purpose of achieving compliance of the work with paragraph E1 of Schedule 1 to the Principal Regulations, he is using one or more design details approved by Robust Details Limited, provided that: (a) the notification specifies: (i) the part or parts of the building in respect of which he is using the design detail;	The requirement for sound-insulation testing came into force on 1 July 2003 for buildings containing rooms for residential purposes. For new houses and buildings containing flats the requirement was deferred until 1 January 2004 to enable the house building industry to prepare a case for substituting the testing requirement with an alternative solution involving the adoption of robust standard details. Ultimately, this date was extended until 1 July 2004 when the new 'Robust Details' were ready to be launched. Therefore, there is an exemption from the need for pre-completion testing for people building new houses or buildings containing flats if they are using design details approved and published by Robust Details Ltd. Unless this is done the only alternative is for pre-completion sound-insulation testing.

Building (Approved Inspectors etc) Regulations 2000	Commentary

(ii) the design detail concerned; and
    (iii) the unique number issued by Robust Details Limited in respect of the specified use of that design detail; and
  (b) the building work carried out in respect of the part or parts of the building identified in the notification is in accordance with the design detail specified in the notification.	

13 Approved inspector's consultation with the fire authority
  (1) In this regulation:
    (a) a 'relevant use' is a use as a workplace of a kind to which Part II of the Fire Precautions (Workplace) Regulations 1997 applies or a use designated under section 1 of the *Fire Precautions Act 1971*;
    (b) a 'relevant building' is a building where it is intended that, after completion of building work, the building or any part of it will be put or will continue to be put to a relevant use;
    (c) a 'relevant change of use' is a material change of use where it is intended that, after the change of use has taken place, the building or any part of it will be put or will continue to be put to a relevant use; and
	Where an initial notice or an amendment notice is to be given (or has been given) in relation to the erection, extension, material alteration or change of use of a building which:
	
	• is to be put to a designated use under the *Fire Precautions Act 1971, section 1*, or
	• will be a workplace subject to Part II of the *Fire Precautions (Workplace) Regulations 1997 (SI 1997/1840)*,
	
	and *Schedule 1, Part B (Fire safety)* of the Principal Regulations also applies, the Approved Inspector is required, before or as soon as practicable after giving the notice, to consult the fire authority. He must give them sufficient plans, and/or other information to show that the work described in the notice will comply with the applicable parts of *Schedule 1, Part B* of the Principal Regulations and must have regard to any views they express.
	    Additionally, before giving a plans certificate or final certificate to the local authority the Approved Inspector must allow the fire authority 15 working days to comment, and have regard to the views they express.

Building (Approved Inspectors etc) Regulations 2000	Commentary
(d) a 'relevant amendment notice' is an amendment notice where any of the work specified in the initial notice, as varied by the amendment notice, being work which could not have been carried out under the original notice ('additional work'), concerns the erection, extension or material alteration of a relevant building or is building work in connection with a relevant change of use of a building and Part B of Schedule 1 to the Principal Regulations imposes requirements in relation to the additional work. (2) This regulation applies where an initial notice is to be given or has been given in relation to the erection, extension or material alteration of a relevant building or in relation to building work in connection with a relevant change of use of a building and Part B of Schedule 1 to the Principal Regulations imposes requirements in relation to the work. (3) Where this regulation applies, the Approved Inspector shall consult the fire authority: (a) before or as soon as practicable after giving an initial notice in relation to the work; (b) before or as soon as practicable after giving a relevant amendment notice in relation to the work; (c) before giving a plans certificate (whether or not combined with an initial notice); and (d) before giving a final certificate.	Some local Acts of Parliament also impose extensive fire authority consultation requirements. The Approved Inspector must undertake any consultation required by local legislation.

Building (Approved Inspectors etc) Regulations 2000	Commentary
(4) Where an Approved Inspector is required by paragraph (3) to consult the fire authority, he shall give to the fire authority: (a) in a case where he is consulting them in connection with an initial notice or an amendment notice, sufficient plans to show whether the work would, if carried out in accordance with those plans, comply with the applicable requirements of Part B of Schedule 1 to the Principal Regulations; and (b) in a case where he is consulting them in connection with the giving of a plans certificate, a copy of the plans in relation to which he intends to give the certificate. (5) Where an Approved Inspector is required by paragraph (3) to consult the fire authority: (a) he shall have regard to any views they express; and (b) he shall not give a plans certificate or a final certificate until 15 days have elapsed from the date on which he consulted them, unless they have expressed their views to him before the expiry of that period. (6) Where a local enactment would, if plans were deposited in accordance with building regulations, require the local authority to consult the fire authority before or during the carrying out of any work, the Approved Inspector shall consult the fire authority in a manner similar to that required by the enactment.	

Building (Approved Inspectors etc) Regulations 2000	Commentary
13A Approved inspector's consultation with the sewerage undertaker (1) This regulation applies where an initial notice or amendment notice is to be given or has been given in respect of work in relation to which paragraph H4 of Schedule 1 to the Principal Regulations imposes requirements. (2) Where this regulation applies, the Approved Inspector shall consult the sewerage undertaker: (a) before or as soon as practicable after giving an initial notice in relation to the work; (b) before or as soon as practicable after giving an amendment notice in relation to the work; (c) before giving a plans certificate (whether or not combined with an initial notice); and (d) before giving a final certificate. (3) Where an Approved Inspector is required by paragraph (2) to consult the sewerage undertaker, he shall give to the sewerage undertaker: (a) in a case where he is consulting them in connection with an initial notice or an amendment notice, sufficient plans to show whether the work would, if carried out in accordance with those plans, comply with the applicable requirements of paragraph H4 of Schedule 1 to the Principal Regulations; and	Where an initial notice or amendment notice is to be given (or has been given) and it is intended to erect, extend or carry out underpinning works to a building within 3 metre of the centreline of a drain, sewer or disposal main to which *Schedule 1 paragraph H4* of the Principal Regulations applies, the Approved Inspector must consult the sewerage undertaker. The procedures and time periods involved parallel to those described for fire authority consultations above.

Building (Approved Inspectors etc) Regulations 2000	Commentary
(b) in a case where he is consulting them in connection with the giving of a plans certificate, a copy of the plans in relation to which he intends to give the certificate. (4) Where an Approved Inspector is required by paragraph (2) to consult the sewerage undertaker: (a) he shall have regard to any views they express; and (b) he shall not give a plans certificate or a final certificate until 15 days have elapsed from the date on which he consulted them, unless they have expressed their views to him before the expiry of that period.	
PART IV: PLANS CERTIFICATES **14 Form of plans certificate** The prescribed form of a plans certificate: (a) which is not combined with an initial notice, shall be Form 3 in Schedule 2; or (b) which is combined with an initial notice, shall be Form 4 in Schedule 2.	The information that must be included in a plans certificate is given in Form 3 of the Building (Approved Inspectors etc) Regulations 2000. A typical certificate is shown in Figure 5.5.
15 Grounds and period for rejecting plans certificate (1) The grounds on which a local authority shall reject a plans certificate which is not combined with an initial notice are those prescribed in Schedule 4. (2) The grounds on which a local authority shall reject a plans certificate combined with an initial notice are those prescribed in Schedule 3 and Schedule 4. (3) The period within which a local authority may give notice of rejection of a plans certificate (whether or not combined with an initial notice) is 5 days beginning on the day on which the certificate is given.	The local authority has 5 working days in which it accept or reject the plans certificate, but it may only reject it on certain specified grounds (see Section 5.6.6). If the plans certificate is combined with an initial notice, the grounds for rejecting an initial notice (see Section 5.6.5) also apply. Plans certificates may be rescinded by a local authority if the work has not started within 3 years of the acceptance date of the certificate.

Building (Approved Inspectors etc) Regulations 2000	Commentary

16 Effect of plans certificate
If an initial notice ceases to be in force as described in section 47(4)(b) of the Act (cancellation etc of initial notice) and the conditions in section 53(2) of the Act (plans certificate given, accepted and not rescinded) are satisfied, the local authority may not:
  (a) give a notice under section 36(1) of the Act (removal or alteration of work which contravenes building regulations); or
  (b) institute proceedings under section 35 of the Act for a contravention of building regulations; in relation to any work described in the certificate which has been carried out in accordance with the plans to which the certificate relates.
	From the client's viewpoint, the advantage of possessing a plans certificate lies in the fact that if at a later stage the initial notice ceases to be effective, the local authority cannot take enforcement action in respect of any work described in the plans certificate if it has been done in accordance with those plans.
	    A plans certificate, when issued by an Approved Inspector, certifies that the design has been checked and that the plans comply with the Building Regulations. Its issue is entirely at the option of the person carrying out the work, and copies of it are sent by the Approved Inspector to that person and to the local authority.

PART V: FINAL CERTIFICATES
 	
17 Form, grounds and period for rejecting final certificate
  (1) The prescribed form of a final certificate shall be Form 5 in Schedule 2 and the grounds on which a local authority shall reject a final certificate are those prescribed in Schedule 5.
  (2) The period within which a local authority may give notice of rejection of a final certificate is 10 days beginning with the day on which the certificate is given.
	The information that must be included in a final certificate is given in Form 5 of the Building (Approved Inspectors etc) Regulations 2000. A typical certificate is shown in Figure 5.6.
	    The local authority has 10 days in which to accept or reject the certificate, but it can only be rejected on certain prescribed grounds. These are described in Section 5.8.4.

PART VI: CESSATION OF EFFECT OF INITIAL NOTICE **18 Events causing initial notice to cease to be in force** (1) Where a final certificate given in respect of work described in an initial notice is rejected, the initial notice shall cease to be in force in relation to the work described in the final certificate on the expiry of a period of 4 weeks beginning with the date on which notice of rejection is given.	If the local authority rejects a final certificate, the initial notice to which it refers will cease to be in force within 4 weeks of the date of rejection. This has the consequences described in regulation 20.

Building (Approved Inspectors etc) Regulations 2000	Commentary

(2) Where work described in an initial notice includes the erection, extension or material alteration of a building, and:
(a) the building or, as the case may be, the extension or any part of the building which has been materially altered is occupied, and
(b) no final certificate is given, the initial notice shall cease to be in force in relation to the building, extension or part of a building which is occupied:
  (i) if the building is to be put to a relevant use as defined by regulation 13(1)(a), on the expiry of a period of 4 weeks beginning with the date of occupation; and
  (ii) in any other case, on the expiry of a period of 8 weeks beginning with the date of occupation.
(3) Where work described in an initial notice involves a material change of use of a building, and:
(a) no final certificate is given, and
(b) that change of use takes place.
The initial notice shall cease to be in force in relation to that change of use on the expiry of a period of 8 weeks beginning with the date on which the change of use takes place.
(4) In any other case where no final certificate is given, an initial notice ceases to be in force on the expiry of a period of 8 weeks beginning with the date on which the work described in the initial notice is substantially completed.
	The client should take great care if it is proposed to occupy a building (or part of a building) before completion. If no final certificate has been given before occupation then the initial notice will cease to have effect on the expiry of certain time periods from the date of occupation, and the work will revert to the local authority. Thus, for most buildings there is a period of grace of 8 weeks from the date of occupation, before the initial notice lapses. However, if the work involves the erection, extension or material alteration of a building and it is to be put to a relevant use as a workplace of a kind to which the *Fire Precautions (Workplace) Regulations 1997, Part II* applies or a use designated under the *Fire Precautions Act 1971, section 1*) the period is reduced to 4 weeks. Currently designated are hotels, boarding houses, offices, shops, railway premises and factories.
	
	If the client needs to occupy part of a building whilst work is continued on the remainder a final certificate may be given for the part to be occupied and further final certificates given as future phases of the work are completed and/or occupied. In this way it is possible to avoid the lapse of the initial notice and the reversion of the work to the local authority. It is of course essential that, at the outset, the client makes the need for such phasing of occupation and completion clear to the Approved Inspector, to avoid later misunderstandings.

Building (Approved Inspectors etc) Regulations 2000	Commentary
(5) An initial notice shall not cease to be in force by virtue of paragraph (2) because part of a building or extension is occupied if a final certificate has been accepted in respect of that part. (6) A local authority may extend any period referred to in this regulation either before or after its expiry.	
19 Cancellation of initial notice (1) Where an Approved Inspector is of the opinion that any of the work described in an initial notice which has been carried out contravenes any provision of building regulations, he may give notice in writing to the person carrying out the work specifying: (a) the requirement of building regulations which in his opinion has not been complied with, and (b) the location of the work which contravenes that requirement. (2) A notice of contravention given in accordance with paragraph (1) shall inform the person carrying out the work that if within the prescribed period he has neither pulled down nor removed the work nor effected such alterations in it as may be necessary to make it comply with building regulations, the Approved Inspector will cancel the initial notice. (3) The period within which the person carrying out the work is to remedy the contravention as described in paragraph (2) is 3 months beginning with the day on which the notice is given.	Approved Inspectors are not empowered to enforce the Building Regulations (this can only be done by local authorities) however, there are powers in regulation 19 which can result in the cancellation of the initial notice and reversion of the work to the local authority, resulting in the resurrection of their full enforcement powers. Reversion of the work to the local authority for enforcement is a final option rarely, if ever, used in practice. It will only arise where the Approved Inspector has tried all other courses of action without success, and feels that the extent of the contravention is so serious that the threat of cancellation of the initial notice is the only option left. In such circumstances, he may give a notice of contravention to the client specifying: • those requirements of the Building Regulations which he feels are not being complied with; and • the location of the non-compliant work.

Building (Approved Inspectors etc) Regulations 2000	Commentary
(4) Form 6 in Schedule 2 is the form of notice to be given by an Approved Inspector to cancel an initial notice in accordance with section 52(1) of the Act in circumstances referred to in section 52(2) of the Act; where notice of a contravention has been given under that subsection and no further initial notice relating to the work has been accepted, that notice shall specify the contravention. (5) Form 7 in Schedule 2 is the form of notice to be given by a person carrying out or intending to carry out work to cancel an initial notice in accordance with section 52(3) of the Act. (6) Form 8 in Schedule 2 is the form of notice to be given by the local authority to cancel an initial notice in accordance with section 52(5) of the Act.	The notice will inform the client that unless the contravention is rectified within a period of 3 months from its date of service, then the initial notice will be cancelled. This gives the client a period of grace in which to instruct the builder to take down and remove the contravening work or to carry out alterations to make it comply. Where the contravening works have not been rectified within the 3-month period the Approved Inspector will cancel the initial notice. This is done by serving a notice of cancellation on the local authority and the client (see Form 6 of the *Building (Approved Inspectors etc) Regulations 2000* as illustrated in Figure 6.1). This notice must give details of the contravention, unless a further initial notice relating to the work has been given and accepted. Full details of enforcement powers are given in Chapter 6. In the following cases, the Approved Inspector must cancel the initial notice by issuing to the local authority a cancellation notice in a prescribed form: • The Approved Inspector has become or expects to become unable to carry out (or continue to carry out) his functions. • The Approved Inspector believes that because of the way in which the work is being carried out he cannot adequately perform his functions. It is also possible for the person carrying out the work to cancel the initial notice. This arises if it becomes apparent that the Approved Inspector is no longer willing or able to carry out his functions (through bankruptcy, death, illness etc). This must be done in the prescribed form and must be served on the local authority and (where practicable), on the Approved Inspector. A typical form is shown in Figure 5.4.

Building (Approved Inspectors etc) Regulations 2000	Commentary
20 Local authority powers in relation to partly completed work (1) This paragraph applies where: (a) any part of the work described in an initial notice has been carried out, (b) the initial notice has ceased to be in force, by reason of regulation 18 or has been cancelled by notice under section 52 of the Act, and (c) no other initial notice relating to that part of the work has been accepted. (2) Where paragraph (1) applies, the owner shall: (a) on being given reasonable notice by the local authority, provide them with: (i) sufficient plans of the work carried out, in respect of which no final certificate has been given, to show whether any part of that work would, if carried out in accordance with the plans, contravene any provision of the Principal Regulations, and (ii) where a plans certificate was given and not rejected in respect of any such part of the work, a copy of the plans to which it relates; and (b) comply with any notice in . writing from the local authority requiring him within a reasonable time to cut into, lay open or pull down so much of the work as prevents the local authority from ascertaining whether any work in relation to which there is no final certificate contravenes any requirement in the Principal Regulations.	Once the initial notice has ceased to have effect, the Approved Inspector will be unable to give a final certificate and the local authority's powers to enforce the Building Regulations can revive unless another Approved Inspector is engaged by the client. If the local authority becomes responsible for enforcing the Regulations it must be provided on request with plans of the building work so far carried out. Additionally, it may require the person carrying out the work to cut into, lay open or pull down work so that it may ascertain whether any work not covered by a final certificate contravenes the Regulations. If it is intended to continue with partially completed work, the local authority must be given sufficient plans to show that the work can be completed without contravention of the Building Regulations. A fee, which is appropriate to that work, will be payable to the local authority.

Building (Approved Inspectors etc) Regulations 2000	Commentary
(3) Where paragraph (1) applies and work in relation to a building has been begun but not completed, a person who intends to carry out further work in relation to the partly completed work shall give the local authority sufficient plans to show that the intended work will not contravene any requirement in the Principal Regulations, including such plans of any part of the work already carried out as may be necessary to show that the intended work can be carried out without contravening any such requirement. (4) Plans given to a local authority in accordance with paragraph (3) are not to be regarded as plans deposited in accordance with building regulations.	
PART VII: PUBLIC BODIES **21 Approval of public bodies** (1) In England if it appears to the Secretary of State, or in Wales, if it appears to the National Assembly for Wales, that: (a) public bodies of a certain description should be enabled to supervise their own work under section 54 of the Act, or (b) that a public body should be approved for the purpose of so supervising its own work, the Secretary of State or, as the case may be, the National Assembly for Wales, shall approve that description of body or, as the case may be, that body in writing and take such steps as appear to them appropriate to inform those local authorities and public bodies which will be affected by the giving of the approval.	Part VII of the 2000 Regulations is concerned with public bodies and, read in conjunction with section 54 and Schedule 4 of the *Building Act 2000*, its effect is to enable designated public bodies to self-certify their own work to which the substantive building regulations apply. Public bodies are approved by the Secretary of State (or in Wales, the National Assembly of Wales), and the regulations relating to notices, consultation with the fire authority and the sewerage undertaker, plans certificates and final certificates mirror those of Part III of the *Building Act* 1984 dealing with Approved Inspectors. To date, only the Metropolitan Police Authority has been prescribed as a public body under the provisions of the *Building Act 1984, section 5*.

Building (Approved Inspectors etc) Regulations 2000	Commentary
(2) In England the Secretary of State, and in Wales, the National Assembly for Wales, may withdraw the approval by a notice in writing given to any public body affected, and shall take such steps as appears to them appropriate to inform local authorities of such withdrawal.	The Secretary of State (or the National Assembly for Wales) may withdraw the approval of a public body by notice in writing.

22 Public body's notice
 (1) the prescribed form of a public body's notice:
   (a) which is not combined with a public body's plans certificate, shall be Form 9 in Schedule 2; or
   (b) which is combined with a public body's plans certificate, shall be Form 11 in Schedule 2.
 (2) A public body's notice shall be accompanied by the plans and documents described in the relevant form prescribed by paragraph (1).
 (3) The grounds on which a local authority shall reject a public body's notice are those prescribed in Schedule 6.
 (4) The period within which a local authority may give notice of rejection of a public body's notice is 10 days beginning with the day on which the notice is given.
	Where a designated public body considers that it can adequately supervise its own building work using its own staff it may serve a public body's notice on the local authority in whose area the work is located. A typical public body's notice is shown in Figure 5.7 which is based on Form 9 from Schedule 2 of the 2000 Regulations.
	
	 The local authority can reject the notice only if it is defective in terms of accuracy and completeness. The grounds for rejection are laid out in Schedule 6 to the 2000 Regulations. Notice of rejection must be given within 10 days of receipt of the notice by the local authority or it is deemed to have been accepted.

23 Public body's consultation with the fire authority Regulation 13 applies where a public body's notice is given as it does where an initial notice is given; and for that purpose there shall be substitutes for references in that regulation to an initial notice, a plans certificate and a final certificate respectively references to a public body's notice, a public body's plans certificate and a public body's final certificate.	The regulations relating to notices, consultation with the fire authority and the sewerage undertaker, plans certificates and final certificates mirror those of Part III of the *Building Act 1984* dealing with Approved Inspectors.

6

Building (Approved Inspectors etc) Regulations 2000	Commentary
23A Public body's consultation with the sewerage undertaker Regulation 13A applies where a public body's notice is given as it does where an initial notice is given; and for that purpose there shall be substitutes for references in that regulation to an initial notice, a plans certificate and a final certificate respectively references to a public body's notice, a public body's plans certificate and a public body's final certificate.	The regulations relating to notices, consultation with the fire authority and the sewerage undertaker, plans certificates and final certificates mirror those of Part III of the *Building Act 1984* dealing with Approved Inspectors.
24 Public body's plans certificate The prescribed form of a public body's plans certificate: (a) which is not combined with a public body's notice, shall be Form 10 in Schedule 2; or (b) which is combined with a public body's notice, shall be Form 11 in Schedule 2.	A public body may give a plans certificate in a similar manner to that for an Approved Inspector. It may also be combined with a public body's notice, at the discretion of the public body. Prescribed forms (Form 10 for a plans certificate and Form 11 for a combined certificate) are presented in Schedule 2 to the 2000 Regulations. The effect of the plans certificate is similar to that for an Approved Inspectors plans certificate described in Section 5.6.7.
25 Grounds and period for rejecting public body's plans certificate (1) The grounds on which a local authority shall reject a public body's plans certificate are those prescribed in Schedule 7. (2) The grounds on which a local authority shall reject a public body's plans certificate combined with a public body's notice are those prescribed in Schedule 6 and Schedule 7. (3) The period within which a local authority may give notice of rejection of a public body's plans certificate or combined notice and certificate is 10 days beginning on the day on which the certificate is given.	The grounds for rejection are given in Schedules 6 and 7. The local authority must reject the public body's notice, combined notice and plans certificate, or plans certificate as appropriate, within 10 days of the notice being given or it is deemed to have been accepted.

Building (Approved Inspectors etc) Regulations 2000	Commentary

26 Effect of public body's plans certificate
If a public body's notice ceases to be in force and the conditions in paragraph 4(2) of Schedule 4 to the Act (public body's plans certificate accepted and not rescinded) are satisfied, the local authority may not:
   (a) give a notice under section 36(1) of the Act (removal or alteration of work which contravenes building regulations); or
   (b) institute proceedings under section 35 of the Act for a contravention of building regulations;
in relation to any work which is described in the certificate and is carried out in accordance with the plans to which the certificate relates.
	The effect of the public body's plans certificate is similar to that for an Approved Inspectors plans certificate described in Section 5.6.6.

27 Public body's final certificate
   (1) The prescribed form of a public body's final certificate shall be Form 12 in Schedule 2 and the grounds on which a local authority shall reject a final certificate are those prescribed in Schedule 8.
   (2) The period within which a local authority may give notice of rejection of a public body's final certificate is 10 days beginning with the day on which the certificate is given.
	The prescribed form for a public body's final certificate is found in Schedule 2 of the 2000 Regulations as Form 12 and the grounds for its rejection are given in Schedule 8. The certificate is similar to an Approved Inspectors final certificate (see Figure 5.6) except that it does not contain a declaration of independence (and therefore the need to declare the existence of minor work) or a declaration that an approved insurance scheme is operative (since none is needed for a public body). The grounds for rejection are similar to those that apply to Approved Inspectors adjusted as above. The local authority must reject the public body's final certificate within 10 days of receipt or it is deemed to have been accepted.

28 Events causing public body's notice to cease to be in force Regulation 18 applies where a public body's notice is given as it does where an initial notice is given; and for that purpose there shall be substitutes for references in that regulation to an initial notice and a final certificate respectively references to a public body's notice and a public body's final certificate.	The events that can cause a public body's notice to cease to be in force are identical to those which apply to an initial notice under the Approved Inspector system and are covered by regulation 18 of the 2000 Regulations (see Section 5.6.5).

Building (Approved Inspectors etc) Regulations 2000	Commentary

PART VIII: CERTIFICATES RELATING TO DEPOSITED PLANS
29 Certificates given under section 16(9) of the Act
(1) Regulations 3 to 7 shall apply in relation to:
   (a) the approval and the termination of approval of persons to certify plans in accordance with section 16(9) of the Act, and
   (b) the designation and the termination of designation of bodies to approve such persons,
as they do in relation to the approval of inspectors and the designation of bodies to approve inspectors respectively.
(2) Regulations 4 and 6 of the Principal Regulations are hereby prescribed for the purposes of section 16(9) of the Act insofar as either requires compliance with:
   (a) Part A (structure) of Schedule 1 to the Principal Regulations, and
   (b) Part L (conservation of fuel and power) of Schedule 1 to the Principal Regulations.
(3) Where deposited plans are accompanied by a certificate as mentioned in section 16(9) of the Act, the evidence of insurance required by that provision is a declaration signed by the insurer that a named scheme of insurance approved by the Secretary of State applies in relation to the certificate which accompanies the plans.
	This regulation deals with the giving of certificates of conformity with building regulations by persons approved for that purpose. Such persons must be approved by a designated body in the same way as for an Approved Inspector. The regulations which apply to the approval and termination of approval of both Approved Inspectors and designated bodies also apply to the approval of approved persons and their designated bodies (see regulations 3 to 7 of the *Building (Approved Inspectors etc) Regulations 2000*). To date, no approved persons have been designated to issue certificates (see also Sections 3.2.4 and 5.2.2).
	So far, regulations 4 and 6 have been prescribed in so far as they relate to Part A (Structure) and Part L (Conservation of fuel and power).

Building (Approved Inspectors etc) Regulations 2000	Commentary
(4) For the purposes of section 16(9) of the Act, the circumstances in which the local authority may reject deposited plans on the grounds referred to in section 16(9)(i) or (ii) are where: (a) the certificate states that the work shown in the plans complies with the requirements of Part A (structure) of Schedule 1 to the Principal Regulations; (b) paragraph A3 of that Schedule applies to the work shown in the plans; and (c) the certificate does not contain a declaration that the person giving the certificate does not, and will not until the work is complete, have a professional or financial interest in the work. (5) The provisions of regulation 10(2) to (5) shall have effect for the purpose of determining whether a person has a professional or financial interest in the work shown in the plans as if references in those provisions to Approved Inspectors were references to persons approved for the purposes of section 16(9) of the Act.	
PART IX: REGISTERS **30 Register of notices and certificates** (1) The register which local authorities shall keep under section 56 of the Act shall contain the information set out in paragraph (2) with respect to: (a) initial notices, amendment notices, notices under section 51C of the Act or public body's notices currently in force, and	Local authorities must keep a register, which is available for inspection by the public at all reasonable times, giving information about notices, certificates, and insurance cover received in connection with their duties under the Approved Inspector system of control described in Sections 5.5 to 5.8.

Building (Approved Inspectors etc) Regulations 2000	Commentary
(b) certificates described in paragraph (3) which have been accepted or are presumed to have been accepted. (2) The information to be registered is: 　(a) the description of the work to which the notice or certificate relates and of the location of the work; 　(b) the name and address of any person who signed the notice or certificate; 　(c) the name and address of the insurer who signed any declaration which accompanied the notice or certificate; and 　(d) the date on which the notice or certificate was accepted or was presumed to have been accepted. (3) The certificates referred to in paragraph (1) are plans certificates, final certificates, public body's plans certificates, public body's final certificates and certificates given under section 16(9) of the Act. (4) A register shall include an index for enabling a person to trace any entry in the register by reference to the address of the land to which the notice or certificate relates. (5) The information prescribed in paragraph (2) shall be entered in the register as soon as practicable and in any event within 14 days of the occurrence to which it relates.	The information must be entered into the register as soon as practicable and not more than 14 days after the event to which it relates. Information about public body's notices and initial notices need only be kept on the register for as long as they are in force. 　The information which must be registered is as follows: • the description and location of the work to which the certificate or notice relates; • the name and address of the person who signed the certificate or notice; • the name and address of the insurer who signed the declaration which accompanied the certificate or notice; and • the date on which the certificate or notice was accepted or was presumed to have been accepted. The register must contain an index which allows any person to trace an entry by reference to the address of the land to which the certificate or notice relates.

Building (Approved Inspectors etc) Regulations 2000	Commentary
PART X: EFFECT OF CONTRAVENING BUILDING REGULATIONS **31 Contravention of certain regulations not to be an offence** Each of these Regulations, other than regulations 12, 12A and 20, is designated as a provision to which section 35 of the Act (penalty for contravening building regulations) does not apply.	In most cases, contravening the *Building (Approved Inspectors etc) Regulations 2000* does not constitute an offence under section 35 of the *Building Act 1984*. The exceptions are in the cases of: • regulation 12 Energy rating (see Section 5.6.6) • regulation 12A Sound-insulation testing (see Section 5.8.2), and • regulation 20 local authority powers in relation to partly completed work (see Sections 5.8.4 and 6.4).
PART XI: MISCELLANEOUS PROVISIONS **32 Transitional provisions** (1) Subject to paragraph (2), the Regulations specified in Schedule 1 shall continue to apply in relation to any building work as if these Regulations had not been made where: (a) before 1 January 2001 an initial notice, an amendment notice or a public body's notice has been given to a local authority; and (b) building work is carried out or is to be carried out on or after that date in accordance with any such notice. (2) Where an initial notice given before 1 January 2001 is varied by an amendment notice given on or after that date, the Regulations specified in Schedule 1 shall continue to apply, as if these Regulations had not been made, in relation to so much of the building work as could have been carried out under that initial notice if the amendment notice had not been given. *Nick Raynsford* Minister of State, Department of the Environment, Transport and the Regions. 13 September 2000.	See commentary to regulation 1(1) above.

SCHEDULE 1: REVOCATION OF REGULATIONS
Regulation 1

Title	Reference	Extent of revocation
The Building (Approved Inspectors etc) Regulations 1985	SI 1985/1066	The whole of the regulations
The Building (Inner London) Regulations 1985	SI 1985/1936	In regulation 2(1) the words 'the Building (Approved Inspectors etc) Regulations 1985' and paragraph 2 of Schedule 2
The Building (Inner London) Regulations 1987	SI 1987/798	In regulation 2(1) the words 'the Building (Approved Inspectors etc) Regulations 1985' and paragraph 2 of Schedule 2
The Building (Amendment) Regulations 1989	SI 1989/1119	Regulation 3
The Building (Approved Inspectors etc) (Amendment) Regulations 1992	SI 1992/740	The whole of the regulations
The Building (Approved Inspectors etc) (Amendment) Regulations 1995	SI 1995/1387	The whole of the regulations
The Building (Approved Inspectors etc) (Amendment) Regulations 1996	SI 1996/1906	The whole of the regulations
The Building (Approved Inspectors etc) (Amendment) Regulations 1998	SI 1998/2332	The whole of the regulations

SCHEDULE 2: FORMS Regulations 8(1), 9(1), 14, 17(1), 19(4), (5) and (6), 22(1), 24 and 27(1) *FORM 1* Section 47 of the Building Act 1984 (the 'Act') The Building (Approved Inspectors etc) Regulations 2000 ('the 2000 Regulations') **Initial Notice**	Form 1 Initial Notice (see Figure 5.3)
FORM 2 Section 51A of the Building Act 1984 (the 'Act') The Building (Approved Inspectors etc) Regulations 2000 ('the 2000 Regulations') **Amendment Notice**	Form 2 Amendment Notice (see Section 5.6.9)
FORM 3 Section 50 of the Building Act 1984 (the 'Act') The Building (Approved Inspectors etc) Regulations 2000 ('the 2000 Regulations') **Plans Certificate**	Form 3 Plans Certificate (see Figure 5.5)
FORM 4 Section 47 and 50 of the Building Act 1984 (the 'Act') The Building (Approved Inspectors etc) Regulations 2000 ('the 2000 Regulations') **Combined Initial Notice and Plans Certificate**	Form 4 Combined Initial Notice and Plans Certificate (see Section 5.6.7)
FORM 5 Section 51 of the Building Act 1984 (the 'Act') The Building (Approved Inspectors etc) Regulations 2000 ('the 2000 Regulations') **Final Certificate**	Form 5 Final Certificate (see Figure 5.6)

FORM 6 Section 52(1) of the Building Act 1984 (the 'Act') The Building (Approved Inspectors etc) Regulations 2000 ('the 2000 Regulations') **Notice of Cancellation by Approved Inspector**	Form 6 Notice of Cancellation by Approved Inspector (see Figure 6.1)
FORM 7 Section 52(3) of the Building Act 1984 (the 'Act') The Building (Approved Inspectors etc) Regulations 2000 ('the 2000 Regulations') **Notice of Cancellation by Person Carrying out the Work**	Form 7 Notice of Cancellation by Person Carrying out Work (see Figure 5.4)
FORM 8 Section 52(5) of the Building Act 1984 (the 'Act') The Building (Approved Inspectors etc) Regulations 2000 ('the 2000 Regulations') **Notice of Cancellation by Local Authority**	Form 8 Notice of Cancellation by Local Authority (see Section 5.6.5)
FORM 9 Section 54 of the Building Act 1984 (the 'Act') The Building (Approved Inspectors etc) Regulations 2000 ('the 2000 Regulations') **Public Body's Notice**	Form 9 Public body's notice (see Figure 5.7)
FORM 10 Paragraph 2 of Schedule 4 to the Building Act 1984 (the 'Act') The Building (Approved Inspectors etc) Regulations 2000 ('the 2000 Regulations') **Public Body's Plans Certificate**	Form 10 Public Body's Plans certificate (see Section 5.9.2)

FORM 11 Paragraph 2(2) of Schedule 4 to the Building Act 1984 (the 'Act') The Building (Approved Inspectors etc) Regulations 2000 ('the 2000 Regulations') **Combined Public Body's Notice and Plans Certificate**	Form 11 Combined Public Body's Notice and Plans Certificate (see Section 5.9.1)
FORM 12 Paragraph 3 of Schedule 4 to the Building Act 1984 (the 'Act') The Building (Approved Inspectors etc) Regulations 2000 ('the 2000 Regulations') **Public Body's Final Certificate**	Form 12 Public Body's Final Certificate (see Section 5.9.3)
SCHEDULE 3: GROUNDS FOR REJECTING AN INITIAL NOTICE, AN AMENDMENT NOTICE, OR A PLANS CERTIFICATE COMBINED WITH AN INITIAL NOTICE Regulations 8(3), 9(3) and 15(2)	See Section 5.6.5
SCHEDULE 4: GROUNDS FOR REJECTING A PLANS CERTIFICATE, OR A PLANS CERTIFICATE COMBINED WITH AN INITIAL NOTICE Regulations 15(1) and 15(2)	See Section 5.6.8
SCHEDULE 5: GROUNDS FOR REJECTING A FINAL CERTIFICATE Regulations 17(1)	See Section 5.8.4
SCHEDULE 6: GROUNDS FOR REJECTING A PUBLIC BODY'S NOTICE, OR A COMBINED PUBLIC BODY'S NOTICE AND PLANS CERTIFICATE Regulations 22(3) and 25(2)	See Sections 5.9.1 and 5.9.2

SCHEDULE 7: GROUNDS FOR REJECTING A PUBLIC BODY'S PLANS CERTIFICATE, OR A COMBINED PUBLIC BODY'S NOTICE AND PLANS CERTIFICATE Regulations 25(1) and 25(2)	See Section 5.9.2
SCHEDULE 8: GROUNDS FOR REJECTING A PUBLIC BODY'S FINAL CERTIFICATE Regulations 27(1)	See Section 5.9.3

Appendix 3

Local Acts of Parliament

Introduction

Where a Local Act is in force, its provisions must also be complied with, since many of these pieces of legislation were enacted to meet local needs and perceived deficiencies in national legislation. The Building Regulations make it clear that local enactments must be taken into account. With the growth and development of Building Regulation control over fire precautions in particular, it is likely that most of the current local legislation is now outdated or has been superseded by the Building Regulations. In fact, some local enactments already contain a statutory bar which gives precedence to Building Regulations. Local authorities are obliged by section 90 of the Building Act to keep a copy of any Local Act provisions and these must be available for public inspection free of charge at all reasonable times.

The following list of Local Acts of Parliament has been updated to reflect the *Building (Repeal of Provisions of Local Acts) Regulations 2003* which came into force on 1 March 2004. The purpose of these regulations was to repeal any parts of Local Acts of Parliament that dealt with separate systems of drainage since these are now covered by Requirement H5 (Separate systems of drainage) in Part H of Schedule 1 to the Building Regulations 2000, which came into force on 1 April 2002. The sections marked with an asterisk (*) in column 3 of the list are applicable where there are matters to be satisfied by a developer either as part of a submission under Building Regulations or which otherwise need to be addressed in parallel with that submission.

County Acts usually apply across a whole county. However, where it is known that a particular section applies only in a particular district this is indicated.

This list was last updated on 9 January 2005:

	(1) **Local Act**	(2) **Relevant sections**	(3)
1	County of Avon Act 1982 Bath City Council	Section 7 Parking places, safety requirements Section 35 Hot springs, excavations in certain areas of Bath	*
2	Berkshire Act 1986	Section 28 Safety of stands Section 32 Access for fire brigade Section 36 Parking places, safety requirements Section 37 Fire precautions in large storage buildings Section 38 Fire precautions in high buildings	 * * * *

3	Bournemouth
	Borough Council
	Act 1985
		Section 15 Access for fire brigade
		Section 16 Parking places, safety requirements
		Section 17 Fire precautions in certain
		        large buildings
		Section 18 Fire precautions in high buildings
		Section 19 Amending section 72
		        Building Act 1984
			*
			*
			*
			
			*

4	Cheshire County Council Act 1980	Section 48 Parking places, safety requirements Section 50 Access for fire brigade Section 49 Fireman switches Section 54 Means of escape, safety requirements	* *
5	County of Cleveland Act 1987	Section 5 Access for fire fighting Section 6 Parking places, safety requirements Section 15 Safety of stands	* *
6	Clwyd Act 1985	Section 19 Parking places, safety requirements Section 20 Access for fire brigade	* *
7	Cornwall Act 1984		
8	Croydon Corporation Act 1960	Section 93/94 Buildings of excess cubic capacity Section 95 Buildings used for trade & for dwellings	*
9	Cumbria Act 1982	Section 23 Parking places, safety requirements Section 25 Access for fire brigade Section 28 Means of escape from certain buildings	* *
10	Derbyshire Act 1981	Section 16 Safety of stands Section 23 Access for fire brigade Section 24 MoE from certain buildings Section 28 Parking places; safety requirements Section 25 Fireman switches	* * * *

	(1) Local Act	(2) Relevant sections	(3)
11	Dyfed Act 1987	Section 46 Safety of stands Section 47 Parking places; safety requirements Section 51 Access for fire brigade	 * *
12	East Ham Corporation Act 1957	Section 54 Separate access to tenements Section 61 Access for fire brigade	* *
13	East Sussex Act 1981	Section 34 Fireman switches Section 35 Access for fire brigade	 *
14	Essex Acts 1952 & 1958 (GLC areas formerly in Essex)		
15	Essex Act 1987	Section 13 Access for fire brigade	*
16	Exeter Act 1987		
17	Greater Manchester Act 1981	Section 58 Safety of stands Section 61 Parking places, safety requirements Section 62 Fireman switches Section 63 Access for fire brigade Section 64 Fire precautions in high buildings Section 65 Fire precautions in large storage buildings Section 66 Fire & safety precautions in public & other buildings	 * * * * *
18	Hampshire Act 1983	Section 11 Parking places, safety requirements Section 12 Access for fire brigade Section 13 Fire precautions in certain large buildings	* * *
19	Hastings Act 1988		
20	Hereford City Council Act 1985	Section 17 Parking places; safety requirements Section 18 Access for fire brigade	* *
21	Humberside Act 1982	Section 12 Parking places, safety requirements Section 13 Fireman switches Section 14 Access for the fire brigade Section 15 Means of escape in certain buildings Section 17 Temporary structures, byelaws	* *
22	Isle of Wight Act 1980 (Part VI)	Section 32 Access for fire brigade Section 31 Fireman switches Section 30 Parking places; safety requirements	* *

	(1) Local Act	(2) Relevant sections	(3)
23	Kent 1958 (GLC areas formerly in Kent)		
24	County of Kent Act 1981	Section 51 Parking places, safety requirements Section 52 Fireman switches Section 53 Access for fire brigade Section 78 Annulment of plans approvals	* * *
25	Lancashire Act 1984	Section 31 Access for fire brigade	*
26	Leicestershire Act 1985	Section 21 Safety of stands Section 49 Parking places, safety requirements Section 50 Access for fire brigade Section 52 Fire precautions in high buildings Section 53 Fire precautions in large storage buildings Section 54 Means of escape, safety requirements	 * * * *
27	County of Merseyside Act 1980	Section 20 Safety of stands Section 48/49 Means of escape from fire Section 50 Parking places, safety requirements Section 51 Fire & safety precautions in public & other buildings Section 52 Fire precautions in high buildings Section 53 Fire precautions in large storage buildings Section 54 Fireman switches Section 55 Access for fire brigade	 * * * * *
28	Middlesex Act 1956 (GLC areas formerly in Middlesex)	Section 33 Access for fire brigade	*
29	Mid Glamorgan County Council Act 1987	Section 9 Access for fire brigade	*
30	Nottinghamshire Act 1985		
31	Plymouth Act 1987		
32	Poole Act 1986	Section 10 Parking places, safety requirements Section 11 Access for fire brigade Section 14 Fire precautions in certain large buildings Section 15 Fire precautions in high buildings	* * *

	(1) **Local Act**	(2) **Relevant sections**	(3)
33	County of South Glamorgan Act 1976	Section 27 Safety of stands Section 48/50 Underground parking places Section 51 Means of escape for certain buildings Section 52 Fireman switches Section 53 Precautions against fire in high buildings Section 54 Byelaws for temporary structures	* *
34	South Yorkshire Act 1980	Section 53 Parking places, safety precautions Section 54 Fireman switches Section 55 Access for fire brigade	* *
35	Staffordshire Act 1983	Section 25 Parking places, safety precautions Section 26 Access for fire brigade	* *
36	Surrey Act 1985	Section 18 Parking places, safety requirements Section 19 Fire precautions in large storage buildings Section 20 Access for fire brigade	* * *
37	Tyne & Wear 1980	Section 24 Access for fire brigade	*
38	West Glamorgan Act 1987	Section 43 Parking places, safety precautions	*
39	West Midlands Act 1980	Section 39 Safety of stands Section 44 Parking places, safety requirements Section 45 Fireman switches Section 46 Access for fire brigade Section 49 Means of escape from certain buildings	 * *
40	West Yorkshire Act 1980	Section 9 Culverting water courses Section 51 Fireman switches	*
41	Worcester City Council Act 1985		

Index